Catalyst Design

Heterogeneous catalysis is widely used in chemical, refinery, and pollution-control processes. For this reason, achieving optimal performance of catalysts is a significant issue for chemical engineers and chemists. This book addresses the question of how catalytic material should be distributed inside a porous support in order to obtain optimal performance. It treats single- and multiple-reaction systems, isothermal and nonisothermal conditions, pellets, monoliths, fixed-bed reactors, and membrane reactors. The effects of physicochemical and operating parameters are analyzed to gain insight into the underlying phenomena governing the performance of optimally designed catalysts. Throughout, the authors offer a balanced treatment of theory and experiment. Particular attention is given to problems of commercial importance. With its thorough treatment of the design, preparation, and utilization of supported catalysts, this book will be a useful resource for graduate students, researchers, and practicing engineers and chemists.

Massimo Morbidelli is Professor of Chemical Reaction Engineering in the Laboratorium für Technische Chemie at ETH, Zürich.

Asterios Gavriilidis is Senior Lecturer in the Department of Chemical Engineering at University College London.

Arvind Varma is the Arthur J. Schmitt Professor in the Department of Chemical Engineering at the University of Notre Dame.

CAMBRIDGE SERIES IN CHEMICAL ENGINEERING

Catalyst Design

OPTIMAL DISTRIBUTION OF CATALYST IN PELLETS, REACTORS, AND MEMBRANES

Massimo Morbidelli
ETH, Zurich

Asterios Gavriilidis
University College London

Arvind Varma
University of Notre Dame

CAMBRIDGE
UNIVERSITY PRESS

To our teachers and students

CAMBRIDGE UNIVERSITY PRESS
Cambridge, New York, Melbourne, Madrid, Cape Town, Singapore, São Paulo

Cambridge University Press
The Edinburgh Building, Cambridge CB2 2RU, UK

Published in the United States of America by Cambridge University Press, New York

www.cambridge.org
Information on this title: www.cambridge.org/9780521660594

First published 2001
This digitally printed first paperback version 2005

A catalogue record for this publication is available from the British Library

Library of Congress Cataloguing in Publication data
Morbidelli, Massimo.
    Catalyst design : optimal distribution of catalyst in pellets, reactors, and membranes /
    Massimo Morbidelli, Asterios Gavriilidis, Arvind Varma.
        p.   cm. – (Cambridge series in chemical engineering)
    Includes bibliographical references.
    ISBN 0-521-66059-9
    1. Catalysts.   I. Gavriilidis, Asterios, 1965–   II. Varma, Arvind.   III. Title.   IV. Series.
TP159.C3 M62   2001
660′.2995 – dc21   00-041460

ISBN-13 978-0-521-66059-4 hardback
ISBN-10 0-521-66059-9 hardback

ISBN-13 978-0-521-01985-9 paperback
ISBN-10 0-521-01985-0 paperback

Contents

Preface

Heterogeneous catalysis is used widely in chemical, refinery and pollution-control processes. Current worldwide catalyst usage is about 10 billion dollars annually, with ca. 3% annual growth rates. While these numbers are impressive, the economic importance of catalysis is far greater since about \$200–\$1,000 worth of products are manufactured for every \$1 worth of catalyst consumed. Further, a vast majority of pollution-control devices, such as catalytic converters for automobiles, are based on catalysis. Thus, heterogeneous catalysis is critically important for the economic and environmental welfare of society.

In most applications, the catalyst is deposited on a high surface area support of pellet or monolith form. The reactants diffuse from the bulk fluid, within the porous network of the support, react at the active catalytic site, and the products diffuse out. The transport resistance of the porous support alters the concentrations of chemical species at the catalyst site, as compared to the bulk fluid. Similarly, owing to heat effects of reaction, temperature gradients also develop between the bulk fluid and the catalyst. The consequence of these concentration and temperature gradients is that reactions occur at different rates, depending on position of the catalyst site within the porous support. In this context, since the catalytic material is often the most expensive component of the catalyst-support structure, the question naturally arises as to how should it be distributed within the support so that the catalyst performance is optimized? This book addresses this question, both theoretically and experimentally, for supported catalysts used in pellets, reactors and membranes.

In Chapter 2, optimization of catalyst distribution in a single pellet is considered, under both isothermal and nonisothermal conditions. Both single and multiple reaction systems following arbitrary kinetics are discussed. Chapter 3 deals with optimization of catalyst distribution in pellets comprising a fixed-bed reactor, while systems involving catalyst deactivation are addressed in Chapter 4. In Chapter 5, the effect of catalyst distribution on the performance of inorganic membrane reactors is presented, where the catalyst is located either in pellets packed inside an inert tubular membrane or within the membrane itself. Issues related to catalysts of significant commercial importance, including automotive, hydrotreating, composite zeolite, biological, and functionalized polymer resin types, are addressed in Chapter 6. The final Chapter 7 considers catalyst preparation by impregnation techniques, where the effects of adsorption, diffusion and drying on obtaining desired nonuniform catalyst distributions within supports are discussed. This book should appeal to all those who are interested in design, preparation and utilization of supported catalysts, including chemical and environmental engineers and chemists. It should also provide a rich source of interesting mathematical problems for applied mathematicians. Finally, we hope that industrial practitioners will find the concepts and results described in this book to be useful for their work.

This book can be used either as text for a senior-graduate level specialized course, or as a supplementary text for existing courses in reaction engineering, industrial chemistry or applied mathematics. It can also be used as a reference for industrial applications.

We thank our departmental colleagues for maintaining an atmosphere conducive to learning. We also thank our families for their encouragement and support, which made this writing possible.

Massimo Morbidelli
Asterios Gavriilidis
Arvind Varma

1

Introduction

1.1 Importance of Catalysis

A large fraction of chemical, refinery, and pollution-control processes involve catalysis. Its importance can be demonstrated by referring to the catalyst market. In 1993 the worldwide catalyst usage was $8.7 billion, comprising $3.1 billion for chemicals, $3 billion for environmental applications, $1.8 billion for petroleum refining, and $0.8 billion for industrial biocatalysts (Schilling, 1994; Thayer, 1994). The total market for chemical catalysts is expected to grow by approximately 20% between 1997 and 2003, primarily through growths in environmental and polymer applications (McCoy, 1999). For the U.S., the total catalyst demand was $2.4 billion in 1995 and is expected to rise to $2.9 billion by the year 2000 (Shelley, 1997). While these figures are impressive, the economic importance of catalysis is even greater when considered in terms of the volume and value of goods produced through catalytic processes. Catalysis is critical in the production of 30 of the top 50 commodity chemicals produced in the U.S., and many of the remaining ones are produced from chemical feedstocks based on catalytic processes. In broader terms, nearly 90% of all U.S. chemical manufacturing processes involve catalysis (Schilling, 1994). Although difficult to estimate, approximately $200–$1000 (Hegedus and Pereira, 1990; Cusumano, 1991) worth of products are manufactured for every $1 worth of catalyst consumed. The value of U.S. goods produced using catalytic processes is estimated to be between 17% and 30% of the U.S. gross national product (Schilling, 1994). In addition, there is the societal benefit of environmental protection, since emission control catalysts are a significant sector of the market (McCoy, 1999).

1.2 Nonuniform Catalyst Distributions

The active materials used as catalysts are often expensive metals, and in order to be utilized effectively, they are dispersed on large-surface-area supports. This approach in many cases introduces intrapellet catalyst concentration gradients during the preparation process, which were initially thought to be detrimental

1

to catalyst performance. The effects of deliberate nonuniform distribution of the catalytic material within the support started receiving attention in the 1960s.

Early publications which demonstrated the superiority of nonuniform catalysts include those of Mars and Gorgels (1964), Michalko (1966a,b), and Kasaoka and Sakata (1968). Mars and Gorgels (1964) showed that catalyst pellets with an inert core can offer superior selectivity during selective hydrogenation of acetylene in the presence of a large excess of ethylene. Michalko (1966a,b) used subsurface-impregnated Pt/Al_2O_3 catalyst pellets for automotive exhaust gas treatment and found that they exhibited better long-term stability than surface-impregnated pellets. Kasaoka and Sakata (1968) derived analytical expressions for the effectiveness factor for an isothermal, first-order reaction with various catalyst activity distributions and showed that those declining towards the slab center gave higher effectiveness factors. A number of publications have dealt with analytical calculations of the effectiveness factor for a variety of catalyst activity distributions. These include papers by Kehoe (1974), Nyström (1978), Ernst and Daugherty (1978), Gottifredi et al. (1981), Lee (1981), Do and Bailey (1982), Do (1984), and Papa and Shah (1992). Some researchers have focused on the issue of shape and activity distribution normalization, where the objective is to provide generalized expressions for the catalytic effectiveness (Wang and Varma, 1978; Yortsos and Tsotsis, 1981, 1982a,b; Morbidelli and Varma, 1983).

Pellets with larger catalyst activity in the interior than on the surface can result in higher effectiveness factors in the case of reactions which behave as negative-order at large reactant concentrations, such as those with bimolecular Langmuir–Hinshelwood kinetics (Villadsen 1976; Becker and Wei, 1977a). Nonuniform catalyst distributions can also improve catalyst performance for reactions following complex kinetics (Juang and Weng, 1983; Johnson and Verykios, 1983, 1984). For example, in multiple-reaction systems, catalyst activity distribution affects selectivity. Shadman-Yazdi and Petersen (1972) and Corbett and Luss (1974) studied an irreversible isothermal first-order consecutive reaction system for a variety of activity profiles. Selectivity to the intermediate species was favored by distributions concentrated towards the external surface of the pellet. Juang and Weng (1983) studied parallel and consecutive reaction systems under nonisothermal conditions. Which catalyst profile amongst those considered gave the best selectivity depended on the characteristics of the particular reaction system. Johnson and Verykios (1983, 1984) and Hanika and Ehlova (1989) studied parallel reaction networks and showed that nonuniform activity distributions can enhance selectivity. Similar improvements were also demonstrated by Cukierman et al. (1983) for the van de Vusse reaction network. Ardiles et al. (1985) considered a bifunctional reacting network representative of hydrocarbon reforming, and showed that selectivity to intermediate products was influenced by the distribution of the two catalytic functions.

The effects of nonuniform activity in catalyst pellets have also been studied in the context of fixed-bed reactors. Minhas and Carberry (1969) studied numerically the advantages of partially impregnated catalysts for SO_2 oxidation in an adiabatic fixed-bed reactor. Smith and Carberry (1975) investigated the production of phthalic anhydride from naphthalene in a nonisothermal nonadiabatic fixed-bed reactor. This is a parallel–consecutive reaction system for which the intermediate product yield is benefited by a pellet with an inert core. Verykios et al. (1983) modeled ethylene epoxidation in a nonisothermal nonadiabatic fixed-bed reactor with nonuniform catalysts. They showed that improved reactor stability against runaway could be obtained, along with higher reactor selectivity and yield, as compared to uniform catalysts.

Rutkin and Petersen (1979) and Ardiles (1986) studied the effect of activity distributions for bifunctional catalysts in fixed-bed reactors, for the case of multiple reaction schemes. Each reaction was assumed to require only one type of catalyst. It was shown that catalyst activity distributions had a strong influence on reactant conversion and product selectivities.

Nonuniform activity distribution for catalysts experiencing deactivation has been studied by a number of investigators (DeLancey, 1973; Shadman-Yazdi and Petersen, 1972; Corbett and Luss, 1974; Becker and Wei, 1977b; Juang and Weng, 1983; Hegedus and McCabe, 1984). If deactivation occurs by sintering, it is minimized by decreasing the local catalyst concentration, i.e., a uniform catalyst offers the best resistance to sintering (Komiyama and Muraki, 1990).

In all cases considered above, catalyst performance was assessed utilizing appropriate indexes. The most common ones include effectiveness, selectivity, yield, and lifetime. Effectiveness factor relates primarily to the reactant conversion that can be achieved by a certain amount of catalyst, while selectivity and yield relate to the production of the desired species in multiple reaction systems. In the case of membrane reactors additional performance indexes (e.g. product purity) become of interest. In deactivating systems, other indexes incorporating the deactivation rate can be utilized apart from catalyst lifetime. Another index, which has not been employed in optimization studies because it is difficult to express in quantitative terms, is attrition. Catalyst pellets with an outer protective layer of support are beneficial in applications where attrition due to abrasion or vibration occurs, since only the inert and inexpensive support is worn off and the precious active materials are retained.

The key parameters which control the effect of nonuniform distribution on the above performance indexes are reaction kinetics, transport properties, operating variables, deactivation mechanism, and catalyst cost. All the early studies discussed above demonstrated that nonuniform catalysts can offer superior conversion, selectivity, durability, and thermal sensitivity characteristics to those wherein the activity is uniform. This was done by comparing the performance of catalysts with selected types of activity profiles, which led to the best profile within the class considered, but not to the optimal one. Morbidelli et al. (1982) first showed that under the constraint of a fixed total amount of active material, the *optimal* catalyst distribution is an appropriately chosen *Dirac-delta function*; i.e., all the active catalyst should be located at a specific position within the pellet. This distribution remains optimal even for the most general case of an arbitrary number of reactions with arbitrary kinetics, occurring in a nonisothermal pellet with finite external heat and mass transfer resistances (Wu et al., 1990a).

It is worth noting that optimization of the catalyst activity distribution is carried out assuming that the support has a certain pore structure and hence specific effective diffusivities for the various components. Thus for a given pore structure, the catalyst distribution within the support is optimized. An alternative optimization in catalyst design is that of pore structure, while maintaining a uniform catalyst distribution. In this case, the mass transport characteristics of the pellet are optimized. This approach has been followed by various investigators and has been shown to lead to improvements in catalyst performance (cf. Hegedus, 1980; Pereira et al., 1988; Hegedus and Pereira, 1990; Beeckman and Hegedus, 1991; Keil and Rieckmann, 1994).

Much effort has also been invested in the preparation of nonuniformly active catalysts. As insight is gained into the phenomena related to catalyst preparation, scientists are able to prepare specific nonuniform profiles. In this regard, it should be recognized that catalyst *loading* and catalyst *activity* distributions are in principle different characteristics. In catalyst preparation, the variable that is usually controlled is the local catalyst loading. However, under reaction conditions, the local reaction rate constant is proportional to catalyst activity. The relation between catalyst activity and catalyst loading is not always straightforward. For structure-sensitive reactions, it depends on the particular reaction system, and hence generalizations cannot be made. On the other hand, for structure-insensitive reactions, catalyst activity is proportional to catalyst surface area. Thus, if the latter depends linearly on catalyst loading, then the catalyst activity and loading distributions are equivalent. If the above dependence is not linear, then the two distributions can be quite different. The majority of studies on nonuniform catalyst distributions address catalyst activity optimization, although a few investigators have considered catalyst loading optimization by postulating some type of surface area–catalyst loading dependence (Cervello et al., 1977; Juang et al., 1981). Along these lines, it was shown that when the relation between catalyst activity and loading is linear, and the latter is constrained by an upper bound, the optimal Dirac-delta distribution becomes a step distribution. However, if this dependence is not linear, which physically means that larger catalyst crystallites are produced with increased loading, then the optimal catalyst distribution is no longer a step, but rather a more disperse distribution (Baratti et al., 1993). An important point is that in order to make meaningful comparisons among various distributions, the total amount of catalyst must be kept constant.

Work in the areas of design, performance, and preparation of nonuniform catalysts has been reviewed by various investigators (Lee and Aris 1985; Komiyama 1985; Dougherty and Verykios 1987; Vayenas and Verykios, 1989; Komiyama and Muraki, 1990; Gavriilidis et al., 1993a). In this monograph, these issues are discussed with emphasis placed on optimally distributed nonuniform catalysts. Special attention is given to applications involving reactions of industrial importance.

1.3   Overview of Book Contents

This book is organized as follows. In Chapter 2, optimization of a single pellet is addressed under isothermal and nonisothermal conditions. Both single and multiple reaction systems are discussed. Starting with simpler cases, the treatment is extended to the most general case of an arbitrary number of reactions with arbitrary kinetics under nonisothermal conditions, in the presence of external transport limitations. The analysis includes the effect of catalyst dispersion varying with catalyst loading. Finally, the improved performance of nonuniform catalysts is demonstrated through experimental studies for oxidation, hydrogenation, and Fischer–Tropsch synthesis reactions.

Optimization of catalyst distribution in pellets constituting a fixed-bed reactor requires one to take into account changes in fluid-phase composition and temperature along the reactor. This is discussed in Chapter 3, for single and multiple reactions, under isothermal and nonisothermal conditions. The discussion of experimental work is focused on catalytic oxidations.

Catalyst distribution influences the performance of systems undergoing deactivation, and this issue is addressed in Chapter 4 for selective as well as nonselective poisoning. Experimental work on methanation, hydrogenation, and NO reduction is presented to demonstrate the advantages of nonuniform catalyst distributions.

In Chapter 5, the effect of catalyst distribution on the performance of inorganic membrane reactors is discussed. In such systems, the catalyst can be located either in pellets packed inside a membrane (IMRCF) or in the membrane itself (CMR). Experimental results for an IMRCF are presented, and the preparation of CMRs with controlled catalyst distribution by sequential slip casting is introduced.

In Chapter 6, special topics of particular industrial importance are discussed. These include automotive catalysts, where various concepts of nonuniform distributions have been utilized; hydrotreating catalysts, which is a particular type of deactivating system; composite catalysts, with more than one catalytic function finding applications in refinery processes; biocatalysts; and functionalized polymer resins, which find applications in acid catalysis.

The final Chapter 7 considers issues related to catalyst preparation. The discussion is focused on impregnation methods, since they represent the most mature technique for preparation of nonuniform catalysts. During pellet impregnation, adsorption and diffusion of the various components within the support are important, and can be manipulated to give rise to desired nonuniform distributions. The chapter concludes with studies where experimental results are compared with model calculations.

2

Optimization of the Catalyst Distribution in a Single Pellet

Among various reaction systems, investigation of optimal catalyst distribution in a *single* pellet has received the most attention. Although the general problem of an arbitrary number of reactions following arbitrary kinetics occurring in a nonisothermal pellet has been solved and will be discussed later in this chapter, it is instructive to first consider simpler cases and proceed gradually to the more complex ones. This allows one to understand the underlying physicochemical principles, without complex mathematical details. Thus, we first treat *single* reactions, under isothermal and nonisothermal conditions, and then analyze *multiple* reactions.

2.1   The Case of a Single Reaction

2.1.1   Isothermal Conditions

In early studies, step distributions of catalyst were analyzed for the simple case of a single reaction occurring under isothermal conditions. Researchers often treated bimolecular Langmuir–Hinshelwood kinetics, which exhibits a maximum in the reaction rate as a function of reactant concentration. Thus, there is a range of reactant concentrations where reaction rate increases as reactant concentration decreases. This feature occurs in many reactions; for example, carbon monoxide or hydrocarbon oxidation, in excess oxygen, over noble metal catalysts (cf. Voltz et al., 1973), acetylene and ethylene hydrogenation over palladium (Schbib et al., 1996), methanation of carbon monoxide over nickel (Van Herwijnen et al., 1973), and water-gas shift over iron-oxide-based catalyst (Podolski and Kim, 1974).

Wei and Becker (1975) and Becker and Wei (1977a) numerically analyzed the effects of four different catalyst distributions. In three of these, the catalyst was deposited in only one-third of the pellet: inner, middle, or outer (alternatively called egg-yolk, egg-white, and eggshell, respectively). In the fourth it was uniformly distributed. The results are shown in Figure 2.1, where the effectiveness factor η is shown as a function of the Thiele modulus ϕ. It may be seen that among these specific distributions, for small values of ϕ (i.e. kinetic control) the inner

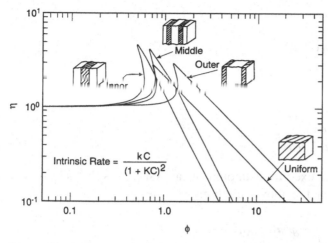

Figure 2.1. Isothermal effectiveness factor η as a function of Thiele modulus ϕ for bimolecular Langmuir–Hinshelwood kinetics in nonuniformly distributed flat-plate catalysts; dimensionless adsorption constant $\sigma = 20$. (From Becker and Wei, 1977a.)

is best, while for large values of ϕ (i.e. diffusion control) the outer is best. For intermediate values of the Thiele modulus, the middle distribution has the highest effectiveness factor. So the question naturally arises: given a Thiele modulus ϕ, among all possible catalyst distributions, which one is the best? This question can be answered precisely, and is addressed next.

Definition of optimization problem

The optimization problem can be stated as follows: given a fixed amount of catalytic material, identify the distribution profile for it within the support which maximizes a given performance index of the catalyst pellet. In order to formulate the problem in mathematical terms, the following equations are required: For a single reaction

$$A \rightarrow \text{products} \tag{2.1}$$

the steady-state mass balance for a single pellet is given by

$$D_e \frac{1}{x^n} \frac{d}{dx}\left(x^n \frac{dC}{dx}\right) = a(x)\,r(C) \tag{2.2}$$

where D_e is the effective diffusivity, x is the space coordinate, C is the reactant concentration, $r(C)$ is the reaction rate, and n is an integer characteristic of the pellet geometry, indicating slab, cylinder, or sphere geometry for $n = 0, 1, 2$ respectively. The catalyst activity distribution function $a(x)$ is defined as the ratio between the local rate constant and its volume-average value:

$$a(x) = k(x)/\bar{k} \tag{2.3}$$

so that by definition

$$\frac{1}{V_p} \int_{V_p} a(x) \, dV_p = 1. \tag{2.4}$$

The boundary conditions (BCs) are

$$x = 0: \qquad \frac{dC}{dx} = 0 \tag{2.5a}$$

$$x = R: \qquad C = C_f. \tag{2.5b}$$

The constraint of a fixed total amount of catalyst means that $\bar{k}V_p$ is constant. In dimensionless form, the above equations become

$$\frac{1}{s^n}\frac{d}{ds}\left(s^n \frac{du}{ds}\right) = \phi^2 a(s) f(u) \tag{2.6}$$

$$s = 0: \qquad \frac{du}{ds} = 0 \tag{2.7a}$$

$$s = 1: \qquad u = 1 \tag{2.7b}$$

$$\int_0^1 a(s)s^n \, ds = \frac{1}{n+1} \tag{2.8}$$

where the following dimensionless quantities have been introduced:

$$u = C/C_f, \qquad s = x/R, \qquad \phi^2 = r(C_f)R^2/D_e C_f.$$
$$f(u) = r(C)/r(C_f) \tag{2.9}$$

Since we are dealing with a single reaction, the catalyst performance is directly related to the *effectiveness factor*, defined by

$$\eta = \frac{\int_0^1 f(u)a(s)s^n \, ds}{\int_0^1 a(s)s^n \, ds} \tag{2.10}$$

which, using equation (2.8), yields

$$\eta = (n+1) \int_0^1 f(u)a(s)s^n \, ds = \frac{n+1}{\phi^2}\left(\frac{du}{ds}\right)_{s=1}. \tag{2.11}$$

Thus, the optimization problem consists in evaluating the catalyst distribution $a(s)$ which maximizes the effectiveness factor η under the constraints given by equations (2.6)–(2.8).

Shape of optimal catalyst distribution
In order to proceed further, we need to know the specific form for the reaction rate $r(C)$. A variety of expressions can be used for this purpose. However, for illustration we choose the bimolecular Langmuir–Hinshelwood kinetics,

$$r(C) = \bar{k}C/(1 + KC)^2 \tag{2.12}$$

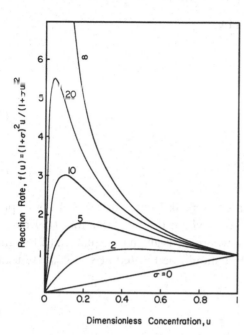

Figure 2.2. Shape of the dimensionless bimolecular Langmuir–Hinshelwood rate function $f(u) = (1+\sigma)^2 u/(1+\sigma u)^2$, for various values of the dimensionless adsorption constant σ. (From Morbidelli et al., 1982.)

so that

$$f(u) = \frac{r(C)}{r(C_f)} = \frac{(1+\sigma)^2 u}{(1+\sigma u)^2} \tag{2.13}$$

where

$$\sigma = KC_f. \tag{2.14}$$

The shape of the rate function $f(u)$ depends on the parameter σ and is shown in Figure 2.2. In particular, $f(u)$ has a unique maximum at

$$u_m = 1/\sigma. \tag{2.15}$$

The dimensionless reaction rate reaches its maximum value in the range $0 < u < 1$ for $\sigma > 1$, and at $u = 1$ for $\sigma \leq 1$. Thus, summarizing, the Langmuir–Hinshelwood kinetics exhibits a maximum value M at $u = u_m$, where

$$u_m = 1/\sigma, \quad M = (1+\sigma)^2/4\sigma \qquad \text{for} \quad \sigma > 1 \tag{2.16a}$$

$$u_m = 1, \qquad M = 1 \qquad\qquad \text{for} \quad \sigma \leq 1 \tag{2.16b}$$

Since $f(u) \leq M$, from the expression for η given by equation (2.11) it is evident that

$$\eta = (n+1) \int_0^1 f(u)a(s)s^n \, ds \leq (n+1)M \int_0^1 a(s)s^n \, ds \tag{2.17}$$

which, using equation (2.8), gives

$$\eta \leq M. \tag{2.18}$$

Therefore, for any activity distribution $a(s)$, the effectiveness factor can never be greater than M. It is apparent that if a function $a(s)$ exists for which $\eta = M$, this will constitute the solution of the optimization problem.

This function exists and is given by

$$a(s) = \frac{\delta(s - \bar{s})}{(n+1)\bar{s}^n} \tag{2.19}$$

where $\delta(s - \bar{s})$ is the Dirac-delta function defined by

$$\delta(s - \bar{s}) = 0 \qquad \text{for all} \quad s \neq \bar{s} \tag{2.20a}$$

and

$$\int_0^1 \delta(s - \bar{s}) \, ds = 1 \tag{2.20b}$$

which physically corresponds to a sharp peak located at $s = \bar{s}$. In our optimization problem, $\bar{s}$ is $\bar{s}_{\text{opt}}$, the value of s where the rate function $f(u)$ reaches its maximum value; i.e., $u(\bar{s}_{\text{opt}}) = u_{\text{m}}$, where u_{m} is given by equation (2.16). In practice, this means that all the catalyst should be located at $s = \bar{s}_{\text{opt}}$. By using the Dirac-delta function property

$$\int_0^1 f(s)\delta(s - \bar{s}) \, ds = f(\bar{s}) \tag{2.21}$$

it can be easily shown that the activity distribution (2.19) is indeed the optimal one. For this, equation (2.19) is substituted into equation (2.11) to give

$$\eta_{\text{opt}} = \int_0^1 f(u) \frac{\delta(s - \bar{s}_{\text{opt}})}{\bar{s}_{\text{opt}}^n} s^n \, ds = f(u_{\text{m}}) = M. \tag{2.22}$$

Evaluation of optimal catalyst location

The evaluation of optimal catalyst location $\bar{s}_{\text{opt}}$ must be performed separately for $\sigma \leq 1$ and $\sigma > 1$.

If $\sigma \leq 1$, then from (2.16b) $u_{\text{m}} = 1$, which is attained at the particle external surface, and hence $\bar{s}_{\text{opt}} = 1$. In this case, from equations (2.16) and (2.22), the effectiveness factor is $\eta_{\text{opt}} = 1$.

If $\sigma > 1$, then some more computations are needed to evaluate the optimal catalyst location. The details are available elsewhere (Morbidelli et al., 1982) and lead to

$$\bar{s}_{\text{opt}} = 1 - \frac{4(\sigma - 1)}{\phi_0^2} \qquad \text{for} \quad n = 0 \tag{2.23a}$$

$$\bar{s}_{\text{opt}} = \exp\left(\frac{8(1 - \sigma)}{\phi_0^2}\right) \qquad \text{for} \quad n = 1 \tag{2.23b}$$

$$\bar{s}_{\text{opt}} = \frac{\phi_0^2}{\phi_0^2 + 12(\sigma - 1)} \qquad \text{for} \quad n = 2 \tag{2.23c}$$

where ϕ_0 is a "clean" Thiele modulus which does not include the adsorption parameter σ and is defined as

$$\phi_0^2 = \bar{k} R^2 / D_{\text{e}} = (1 + \sigma)^2 \phi^2. \tag{2.24}$$

Table 2.1. Location of optimal catalyst distribution and corresponding effectiveness factor for isothermal, bimolecular Langmuir–Hinshelwood kinetics, without the existence of external transport resistances.

	$\Lambda \leq 1$	$\sigma > 1$
$\bar{s}_{opt}$	1	$1 - \Lambda^a \quad (n = 0)$
		$\exp(-\Lambda) \quad (n = 1)$
		$\dfrac{1}{1 + \Lambda} \quad (n = 2)$
η_{opt}	1	$\dfrac{(1 + \sigma)^2}{4\sigma}$

^{a} For $\Lambda > 1$, one has $\bar{s}_{opt} = 0$ and $\eta_{opt} = f(\bar{u})$, where $\bar{u}$ is the solution of $1 - \bar{u} - \phi^2 f(\bar{u}) = 0$. Here $f(\bar{u})$ is given by equation (2.13).

The corresponding optimal effectiveness factor is

$$\eta_{opt} = \frac{(1 + \sigma)^2}{4\sigma}. \tag{2.25}$$

The effect of all the involved physicochemical parameters on $\bar{s}_{opt}$ can be expressed through a single dimensionless parameter Λ, which is used in Table 2.1 to summarize the results obtained so far:

$$\Lambda = \frac{4(n + 1)(\sigma - 1)}{\phi_0^2}. \tag{2.26}$$

From inspection of equation 2.23 it appears that while for the infinite cylinder ($n = 1$) and the sphere ($n = 2$) we have $0 \leq \bar{s}_{opt} \leq 1$ for all values of $\phi_0 \geq 0$ and $\sigma \geq 1$, for the infinite slab ($n = 0$) the value of $\bar{s}_{opt}$ can become negative. This is physically unrealistic, and in this case the optimal catalyst distribution is $\bar{s}_{opt} = 0$, i.e., the catalyst must be concentrated at the pellet center (Morbidelli et al., 1982). The resulting value of the effectiveness factor will in this case be smaller than η_{opt} given by equation (2.25), but it is still the maximum obtainable for the given values of ϕ_0 and σ.

From equations (2.23), it is seen that the catalyst location depends on the physicochemical parameters of the system, i.e., the Thiele modulus ϕ_0 and the adsorption constant σ. On increasing the Thiele modulus or decreasing the adsorption constant, the optimal location of the active catalyst moves from the interior of the pellet to the external surface. This is shown in Figure 2.3, where for the spherical pellet the $\bar{s}_{opt}$-vs-ϕ_0 curves for various σ values are plotted. Increasing the Thiele modulus (keeping the adsorption constant unchanged) leads to larger diffusional resistances, and therefore moves the location where $u = u_m$ closer to the pellet's external surface. Similarly, decreasing the adsorption constant (keeping the Thiele modulus unchanged) causes the maximum of the reaction rate to occur at larger u values, which again are encountered closer to the external surface.

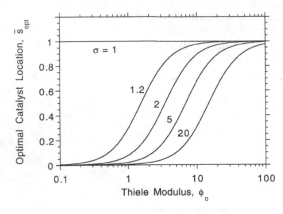

Figure 2.3. Optimal catalyst location $\bar{s}_{opt}$ as a function of Thiele modulus ϕ_0 for various values of the dimensionless adsorption constant σ. (From Morbidelli et al., 1982.)

Step catalyst distribution

From the practical point of view it is not possible, and from the sintering point of view it is not desirable, to locate the catalyst as a Dirac-delta distribution. The question therefore arises if the Dirac-delta distribution can be approximated by a more convenient step-type distribution of narrow width. In this case, the catalyst distribution is described by

$$a(s) = \begin{cases} 0 & \text{for} \quad s < s_1 \text{ or } s > s_2 \\ a & \text{for} \quad s_1 < s < s_2 \end{cases} \tag{2.27}$$

where $s_1 = \bar{s} - \Delta$, $s_2 = \bar{s} + \Delta$, and a is a constant which is evaluated using condition (2.8) as follows:

$$a = \frac{1}{s_2^{n+1} - s_1^{n+1}}. \tag{2.28}$$

The system of equations (2.6)–(2.7) for the step distribution can only be solved numerically. Care must be exercised in finding all possible solutions, since multiple solutions may exist.

In Figure 2.4, the effectiveness factor is plotted as a function of Thiele modulus for given values of σ, $\bar{s}$, and Δ. Note that the maximum value of the effectiveness factor η_m is attained in a region of ϕ where multiplicity is present. This η_m value is shown as a function of the step-distribution half thickness in Figure 2.5. As the step width decreases, the behavior of the step distribution approaches that of the

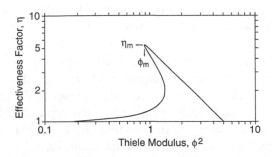

Figure 2.4. Effectiveness factor η vs Thiele modulus ϕ^2 for a slab pellet with step distribution of catalyst centered at $\bar{s} = 0.8$, and half thickness $\Delta = 0.01$. Bimolecular Langmuir–Hinshelwood reaction, $\sigma = 20$. Here η_m is the maximum value of the effectiveness factor, and ϕ_m the corresponding Thiele modulus. (From Morbidelli et al., 1982.)

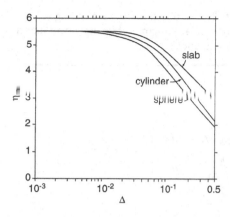

Figure 2.5. Maximum value of effectiveness factor η_m for a step distribution of catalyst centered at $\bar{s} = 0.5$, as a function of the step width Δ. Bimolecular Langmuir–Hinshelwood reaction, $\sigma = 20$. (From Morbidelli et al., 1982.)

corresponding Dirac-delta distribution, i.e., as $\Delta \to 0$, $\eta_m \to \eta_{opt} = 5.51$ as given by equation (2.25). For the parameter values considered in Figure 2.5, it can be seen that if the thickness of the active layer is less than ~5% of the pellet characteristic dimension, the behavior of the two distributions is virtually the same. It is worth noting that such widths can be realized experimentally, as discussed in section 7.2.

Sensitivity of catalyst performance to the step location

The sensitivity of catalyst performance to the step distribution location is of great importance, due to the inevitable difficulties encountered experimentally in locating the catalyst at a precise point $\bar{s}_{opt}$. This is illustrated in Figure 2.6, where the effectiveness factor is plotted as a function of the location of the active zone center.

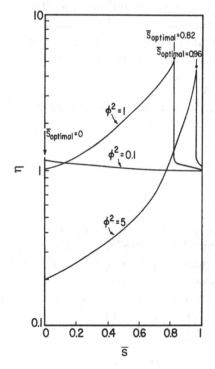

Figure 2.6. Values of the effectiveness factor η for a pellet with step distribution of catalyst centered at location $\bar{s}$. Each curve is characterized by a value of the Thiele modulus ϕ and by the corresponding optimal location $\bar{s}_{opt}$. Bimolecular Langmuir–Hinshelwood reaction, $n = 0$, $\sigma = 20$, $\Delta = 0.025$. (From Morbidelli et al., 1982.)

Table 2.2. Location of optimal catalyst distribution and corresponding effectiveness factor for isothermal, bimolecular Langmuir–Hinshelwood kinetics in the presence of external transport resistances.

	$\sigma \leq 1$	$\sigma > 1$	
		$\mathbf{Bi_m > Bi^*}$	$\mathbf{Bi_m < Bi^*}$
$\bar{s}_{opt}$	1	$1 - \Lambda_e$ [a] $n=0$ $\exp(-\Lambda_e)$ $n=1$ $\dfrac{1}{1+\Lambda_e}$ $n=2$	1
η_{opt}	$f(\bar{u})$	$\dfrac{(1+\sigma)^2}{4\sigma}$	$f(\bar{u})$
	with $\bar{u}$ given by		with $\bar{u}$ given by
	$1 - \bar{u} - \dfrac{\phi^2 f(\bar{u})}{Bi_m(n+1)} = 0$		$1 - \bar{u} - \dfrac{\phi^2 f(\bar{u})}{Bi_m(n+1)} = 0$

[a] For $Bi_m > Bi^*/(1 - Bi^*)$ and $Bi^* < 1$ we have $\bar{s}_{opt} = 0$, $\eta_{opt} = f(\bar{u})$, and $\bar{u}$ is given by $1 - \bar{u} - [(1 + Bi_m)/Bi_m]\phi^2 f(\bar{u}) = 0$. Here $f(\bar{u})$ is given by equation (2.13).

If this is at $\bar{s} < \bar{s}_{opt}$, then the effectiveness factor is lower than its maximum value to an extent proportional to the difference $\bar{s} - \bar{s}_{opt}$. However, if the location of the active layer is $\bar{s} > \bar{s}_{opt}$, then the effectiveness factor undergoes a dramatic decrease. The reason is related to the pellet's multiplicity behavior. As illustrated in Figure 2.4, a certain level of diffusional resistance is required to give the maximum value of the effectiveness factor. If the actual diffusional resistance is larger, i.e. $\phi > \phi_m$, the effectiveness factor on the upper branch undergoes a slight decrease, while if it is smaller, i.e. $\phi < \phi_m$, it undergoes a dramatic decrease, since the upper branch is no longer available. Thus, in preparing catalysts, if an error is anticipated, it is better to err towards placing the step deeper within the pellet than $\bar{s}_{opt}$.

Effect of external mass transfer resistance

In the presence of external mass transfer resistance, the pellet surface BC (2.7b) changes to

$$s = 1: \qquad \frac{du}{ds} = Bi_m(1 - u) \tag{2.29}$$

where

$$Bi_m = k_g R / D_e \tag{2.30}$$

is the mass Biot number. In this case, the optimal activity distribution is again given by a Dirac-delta function (Morbidelli and Varma, 1982). The effect of the external mass transfer resistance on the optimal location is similar to that of internal resistance. In particular, for decreasing values of Bi_m, the optimal location moves towards the external surface of the pellet. It can be obtained from the equations reported in Table 2.2, where the parameter Λ_e is given by

$$\Lambda_e = -\frac{1}{Bi_m} + \frac{1}{Bi^*} \tag{2.31}$$

where

$$Bi^* = \frac{\phi_0^2}{4(n+1)(\sigma-1)}.$$
(2.32)

2.1.2 Nonisothermal Conditions

When the heat of reaction is high, the assumption of isothermality is not valid and significant temperature gradients may develop. For typical gas–solid catalytic reactions, temperature gradients within the pellet are negligible in comparison with those between the bulk fluid and external pellet surface (cf. Carberry, 1976). However, for pellets with nonuniform activity distribution, internal temperature gradients may become important. This has been demonstrated by Butt and coworkers for nonuniform pellets resulting from deactivation (Kehoe and Butt, 1972; Butt et al., 1977; Lee et al., 1978; Downing et al., 1979).

Definition of optimization problem

The case of arbitrary reaction kinetics will be taken up later, in section 2.1.3. However, to illustrate the effect of temperature gradients, let us continue to consider a single bimolecular Langmuir–Hinshelwood reaction (cf. Morbidelli et al., 1985),

$$A \rightarrow \text{products.}$$
(2.33)

The steady-state mass and energy balances within the pellet under nonisothermal conditions are

$$D_e \frac{1}{x^n} \frac{d}{dx} \left(x^n \frac{dC}{dx} \right) = a(x) r(C, T)$$
(2.34)

$$\lambda_e \frac{1}{x^n} \frac{d}{dx} \left(x^n \frac{dT}{dx} \right) = -(-\Delta H) a(x) r(C, T)$$
(2.35)

where λ_e is the effective thermal conductivity, T is the temperature, and $-\Delta H$ is the heat of reaction, while all the other symbols have been defined earlier.

The BCs are

$$x = 0: \qquad \frac{dC}{dx} = 0, \quad \frac{dT}{dx} = 0$$
(2.36a)

$$x = R: \qquad C = C_f, \quad T = T_f.$$
(2.36b)

The catalyst distribution $a(x)$ for nonisothermal systems is defined as the ratio between the local rate constant and its volume-average value, both evaluated at the bulk fluid temperature:

$$a(x) = k(x, T_f)/\bar{k}(T_f).$$
(2.37)

It is thus independent of temperature. In dimensionless form, the above equations become

$$\frac{1}{s^n}\frac{d}{ds}\left(s^n\frac{du}{ds}\right) = \phi^2 a(s) f(u,\theta) \tag{2.38}$$

$$\frac{1}{s^n}\frac{d}{ds}\left(s^n\frac{d\theta}{ds}\right) = -\beta\phi^2 a(s) f(u,\theta) \tag{2.39}$$

$$s = 0: \qquad \frac{du}{ds} = 0, \qquad \frac{d\theta}{ds} = 0 \tag{2.40a}$$

$$s = 1: \qquad u = 1, \quad \theta = 1 \tag{2.40b}$$

where

$$u = \frac{C}{C_f}, \qquad s = \frac{x}{R}, \qquad \phi^2 = \frac{r(C_f, T_f) R^2}{D_e C_f}, \qquad f(u,\theta) = \frac{r(C,T)}{r(C_f, T_f)}$$

$$\theta = \frac{T}{T_f}, \qquad \beta = \frac{(-\Delta H) D_e C_f}{\lambda_e T_f}. \tag{2.41}$$

Combining equations (2.38) and (2.39), the following invariance can be obtained (cf. Aris, 1975):

$$\frac{d}{ds}\left(s^n\frac{d}{ds}(\theta + \beta u)\right) = 0 \tag{2.42}$$

Using the above relationship and the BCs (2.40), the dimensionless temperature θ can be expressed as a function of dimensionless concentration u as follows:

$$\theta = 1 + \beta(1 - u). \tag{2.43}$$

This allows us to reduce the problem to a single differential equation

$$\frac{1}{s^n}\frac{d}{ds}\left(s^n\frac{du}{ds}\right) = \phi^2 a(s) f[u, 1 + \beta(1 - u)]. \tag{2.44}$$

Shape of optimal catalyst distribution
Equation (2.44) is the same as equation (2.6); therefore, following the same procedure as in the isothermal case, it can be proven that the optimal distribution which maximizes the effectiveness factor is again a Dirac-delta function (Morbidelli et al., 1985). The only difference is that the dimensionless reaction rate now is

$$f(u) = \frac{(1+\sigma)^2 u \exp\left(\dfrac{\gamma\beta(1-u)}{1+\beta(1-u)}\right)}{\left[1 + \sigma u \exp\left(-\dfrac{\varepsilon\beta(1-u)}{1+\beta(1-u)}\right)\right]^2} \tag{2.45}$$

where three new parameters appear: the dimensionless heat of reaction β, and the activation energies γ and ε for the reaction rate constant and the adsorption constant, given by

$$\gamma = \frac{E}{R_g T_f}, \qquad \varepsilon = \frac{-\Delta H_a}{R_g T_f}. \tag{2.46}$$

Table 2.3. Location of optimal catalyst distribution and corresponding effectiveness factor for nonisothermal, bimolecular Langmuir–Hinshelwood kinetics, without the existence of external transport resistances.

	$\sigma \leq \sigma_c$	$\sigma > \sigma_c$	
$\bar{s}_{opt}$	1	$1 - \Lambda'^a$	$n = 0$
		$\exp(-\Lambda')$	$n = 1$
		$\dfrac{1}{1 + \Lambda'}$	$n = 2$
η_{opt}	1	$f(u_m)$,	
		where u_m is given by $f'(u_m) = 0$	

^{a} For $\Lambda' > 1$, one has $\bar{s}_{opt} = 0$ and $\eta_{opt} = f(\bar{u})$, where $\bar{u}$ is the solution of $1 - \bar{u} - \phi^2 f(\bar{u}) = 0$. Here $f(\bar{u})$ is given by equation (2.45), and σ_c is given by equation (2.47).

Evaluation of optimal catalyst location

The optimal performance of the catalyst pellet is obtained by locating the Dirac-delta catalyst distribution so as to maximize the reaction rate by taking advantage of both temperature and concentration gradients inside the pellet. Higher temperature increases the reaction rate through a higher reaction rate constant, while higher reactant concentration can either increase or decrease the rate of reaction, depending upon the extent of reactant inhibition (see Figure 2.2).

The expressions for the optimal location and the corresponding effectiveness factor are summarized in Table 2.3, where the critical value σ_c, which determines if the optimal catalyst distribution is a surface or a subsurface one, is given by

$$\sigma_c = \frac{1 - \beta\gamma}{1 + \beta(\gamma + 2\varepsilon)}. \tag{2.47}$$

The effects of all physicochemical variables are described through a single dimensionless parameter

$$\Lambda' = \frac{(n + 1)(1 - u_m)}{\phi^2 f(u_m)} \tag{2.48}$$

where u_m is the value of u which maximizes the reaction rate $f(u)$ given by equation (2.45).

In general, a larger dimensionless heat of reaction β leads to higher temperatures within the pellet, thus increasing the ratio between reaction and diffusion rates. This makes the reactant concentration profile steeper and therefore makes the optimal location approach the external surface of the pellet (see Figure 2.7). However, for small values of σ the optimal location can exhibit a minimum as a function of β, as illustrated in Figure 2.7. As β increases, the optimal location moves from the surface (at $\beta = 0$) towards the pellet interior, reaches a minimum value, and then moves back to the pellet surface ($\beta \to \infty$).

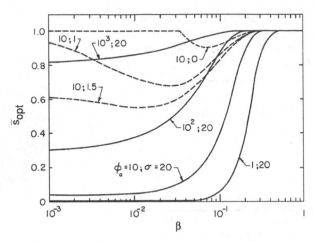

Figure 2.7. Optimal catalyst location $\bar{s}_{\text{opt}}$ as a function of the heat of reaction parameter β for various ϕ_0 and σ values. Bimolecular Langmuir–Hinshelwood reaction, $n = 2$, $\gamma = 30$, $\varepsilon = 2.33$. (From Morbidelli et al., 1985.)

Note that $\sigma = 0$ corresponds to a *first-order* reaction. In this case, under *isothermal* conditions, the reaction rate is maximized at the location where the reactant concentration is the highest, i.e. at the external surface of the pellet. However, if the pellet is *nonisothermal*, the optimal location can well be in the interior of the pellet for a range of β values (see curve corresponding to $\sigma = 0$ in Figure 2.7). More specifically, subsurface optimal catalyst locations are attained when $\sigma > \sigma_c$ (see Table 2.3), which for a first-order reaction leads to

$$\beta\gamma > 1. \tag{2.49}$$

This behavior is solely due to the intrapellet temperature gradient, and indicates that the temperature increase at the location of the reaction causes the net reaction rate to increase sufficiently so as to more than overcome the adverse effect of its decrease due to decreased reactant concentration. Due to the above temperature gradient, effectiveness factors larger than unity can be attained.

2.1.3   Arbitrary Kinetics with External Transport Resistances

In the cases discussed above in sections 2.1.1 and 2.1.2, a single-variable (viz. reactant concentration) optimization problem was solved. This strategy continues to work for the case of a single reaction following arbitrary kinetics, as long as either there are no external heat and mass transport resistances ($Bi_m = Bi_h = \infty$), or with $Bi_m = Bi_h$ when the resistances are finite. However, when $Bi_m \neq Bi_h$, a two-variable optimization problem has to be solved. This problem has been addressed through different approaches by Chemburkar et al. (1987) and by Vayenas and Pavlou (1987a). Both studies showed that a Dirac-delta distribution remains the optimal one. In this case the differential equations are reduced to a single equation but with two variables as illustrated below.

Evaluation of optimal catalyst location

The steady-state mass and energy balances within the pellet are again given by equations (2.34) and (2.35). The BC at the pellet center also remains the same as equation (2.36a). However, owing to the finite external mass and heat transport resistances, the pellet surface BC (2.36b) now takes the form

$$x = R: \qquad D_e \frac{dC}{dx} = k_g(C_f - C), \qquad \lambda_e \frac{dT}{dx} = h(T_f - T). \qquad (2.50)$$

The corresponding dimensionless equations are (2.38), (2.39), (2.40a), and

$$s = 1: \qquad \frac{du}{ds} = \text{Bi}_m(1 - u), \qquad \frac{d\theta}{ds} = \text{Bi}_h(1 - \theta) \qquad (2.51)$$

where

$$\text{Bi}_m = k_g R/D_e, \qquad \text{Bi}_h = hR/\lambda_e. \qquad (2.52)$$

From equations (2.38)–(2.40a) and (2.51), a relation between the dimensionless concentration u and temperature θ at any location s can be obtained through standard analysis (Aris, 1975). This gives

$$\theta = 1 + \beta\mu + \beta u_s(1 - \mu) - \beta u \qquad (2.53)$$

where $\mu = \text{Bi}_m/\text{Bi}_h$ and the subscript s refers to the value at the pellet external surface, i.e. $s = 1$. Using equation (2.53), the system of two differential equations with variables u and θ reduces to one differential equation

$$\frac{1}{s^n} \frac{d}{ds} \left(s^n \frac{du}{ds} \right) = \phi^2 a(s) d(u, u_s) \qquad (2.54)$$

where $d(u, u_s)$ is given by $f(u, \theta)$ with θ replaced by (2.53). Note that two variables u and u_s are present, since the value of u_s is not known *a priori*. For this reason, the procedure used in previous cases (sections 2.1.1 and 2.1.2) cannot be applied here. Details of the original analysis by Chemburkar et al. (1987) are not presented here, because the simpler and more general proof in section 2.3.2 can also be applied to this case.

Let us now simply report the final results. The optimal catalyst location is given by

$$\bar{s}_{\text{opt}} = 1 - \Lambda'_e \qquad \text{for} \quad n = 0 \qquad (2.55a)$$
$$\bar{s}_{\text{opt}} = \exp(-\Lambda'_e) \qquad \text{for} \quad n = 1 \qquad (2.55b)$$
$$\bar{s}_{\text{opt}} = 1/(1 + \Lambda'_e) \qquad \text{for} \quad n = 2 \qquad (2.55c)$$

where

$$\Lambda'_e = \frac{1}{\text{Bi}_m} \cdot \frac{u_s^\delta - u_m}{1 - u_s^\delta}. \qquad (2.56)$$

Note that Λ'_e is always positive, so that equations (2.55) always provide physically meaningful values for $\bar{s}_{\text{opt}}$, i.e. $0 < \bar{s}_{\text{opt}} < 1$. The only exception is for the slab geometry ($n = 0$), where Λ'_e can be larger than 1, in which case $\bar{s}_{opt} = 0$. In order to use the relation above we need to compute u_m and u_s^δ. The first, u_m, is the value of u which maximizes the function $d(u, u_s^\delta)$, while u_s^δ is the smallest root of the equation

$$\frac{(n+1)\text{Bi}_m}{\phi^2}\left(1 - u_s^\delta\right) = d\left(u_m, u_s^\delta\right). \tag{2.57}$$

This also provides the value of the optimal effectiveness factor, given by

$$\eta_{opt} = d\left(u_m, u_s^\delta\right). \tag{2.58}$$

The number of parameters upon which the active layer location depends is now larger, and for the case of a nonisothermal mth-order reaction these are the dimensionless heat of reaction β, the dimensionless activation energy γ, the reaction order m, the geometry integer n, the mass Biot number Bi_m, the ratio μ of mass to heat Biot numbers, and the Thiele modulus ϕ. These dimensionless parameters in turn contain various physicochemical variables. However, given a reaction, a fixed amount of catalyst, and a support, only two of them can be considered as operating variables, viz. T_f and C_f. Any change in them affects ϕ, β and γ, although the effect is not substantial for the latter two. Hence, the dependence of the optimal location $\bar{s}_{opt}$ on the Thiele modulus ϕ is shown in Figure 2.8 for typical values of the other parameters while one of them is varied around its base value. The optimal catalyst activity location $\bar{s}_{opt}$ can exhibit a discontinuity as a specific physicochemical parameter is varied, and this arises from the fact that often multiple steady states are possible.

The effect of external transport resistances is illustrated in Figure 2.8.a and b. As expected, when the resistances increase, i.e. when the mass or heat Biot number decreases, the optimal location moves outwards. Also, the discontinuity in the $\bar{s}_{opt}$-vs-ϕ plot occurs at lower values of ϕ as the transport resistances increase. The effect of pellet geometry (Figure 2.8.c) is that as n increases, the jump occurs at a higher value of ϕ. It can also be seen in Figure 2.8.d–f that in general as β increases (or γ increases, or m decreases): (1) the optimal location moves inward for values of ϕ sufficiently small that no jump has occurred, and (2) the jump occurs at smaller ϕ. However, for sufficiently small values of β or γ (or sufficiently large values of m), the optimal location $\bar{s}_{opt}$ is always 1, regardless of the value of the Thiele modulus. This behavior is directly related to the fact that when

$$\frac{\beta\gamma}{m} < 1 \tag{2.59}$$

for finite Bi_m and Bi_h, the optimal location for all ϕ is $\bar{s}_{opt} = 1$ [compare with equation (2.49)].

The discontinuity that the optimal catalyst location exhibits as a physicochemical parameter is varied results in a similar discontinuity in the effectiveness factor

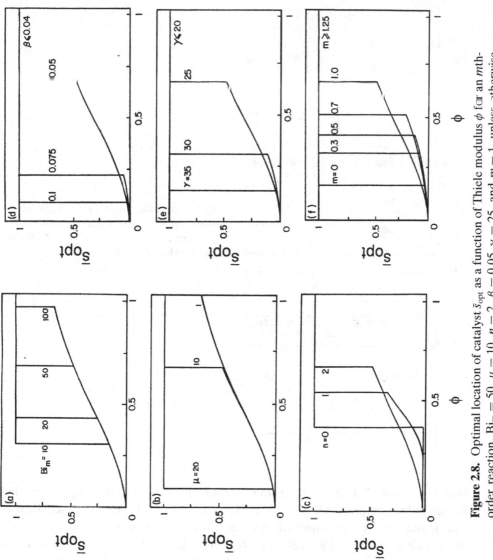

Figure 2.8. Optimal location of catalyst $\bar{s}_{opt}$ as a function of Thiele modulus ϕ for an mth-order reaction. $Bi_m = 50$, $\mu = 10$, $n = 2$, $\beta = 0.05$, $\gamma = 25$, and $m = 1$, unless otherwise specified. (From Chemburkar et al., 1987.)

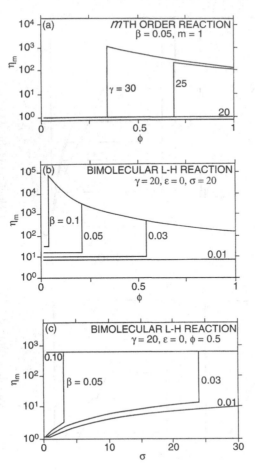

Figure 2.9. Effect of ϕ or σ on the maximum value of the effectiveness factor η_m. Here $Bi_m = 50$, $\mu = 10$, $n = 2$. (From Chemburkar et al., 1987.)

(see Figure 2.9). When the jump occurs from a lower to a higher $\bar{s}_{opt}$ value, the higher value is typically 1, i.e. at the external surface of the pellet (see Figure 2.8.e), and the corresponding effectiveness factor also jumps by some orders of magnitude (see Figure 2.9.a). So, if there is free choice of the operating variables (particularly the bulk fluid values T_f and C_f), then the optimal catalyst location can always be moved to the external surface of the pellet. In addition to the convenience of preparing eggshell catalysts, such a strategy will also result in an increased catalytic effectiveness.

Sensitivity of catalyst performance to catalyst location

Figure 2.9.a,b shows that there is a relatively narrow range of ϕ that gives a tremendous enhancement in the effectiveness factor, and in the case of any deviations from the set ϕ value, it is safer to have deviations leading to higher ϕ. Note that Thiele modulus can be changed conveniently, for example by changing the bulk fluid temperature T_f. The sensitivity of the effectiveness factor to ϕ is demonstrated in Figure 2.10. For the given set of parameter values and $\phi = 0.56$, the optimum catalyst location is $\bar{s}_{opt} = 0.5$. However, for this $\bar{s}$ even a slight decrease

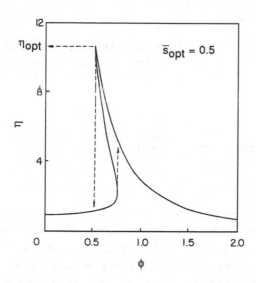

Figure 2.10. Effectiveness factor η of a pellet with a given Dirac-delta activity distribution as a function of the Thiele modulus ϕ. Bimolecular Langmuir–Hinshelwood reaction, $Bi_m = 50$, $\mu = 10$, $n = 2$, $\beta = 0.03$, $\gamma = 20$, $\varepsilon = 0$, $\sigma = 20$. (From Chemburkar et al., 1987.)

in the operating ϕ can cause a catastrophic decrease in the effectiveness factor. Thus, from the practical point of view, it would be desirable to operate at a ϕ somewhat higher than 0.56, giving a somewhat lower effectiveness factor on the upper branch, but for which any small deviations from the operating ϕ on either side do not cause a large deviation in the obtained effectiveness factor. This result is similar to that discussed previously in the context of an isothermal reaction (cf. Figures 2.4 and 2.6).

For bimolecular Langmuir–Hinshelwood reactions, the operating variables T_f and C_f have significant effect not only on ϕ but also on σ. As shown in Figure 2.9.c, there is no decrease in effectiveness factor when σ is increased beyond the value at which the jump occurs. This indicates that the effectiveness factor is insensitive to changes in σ, as long as the operating σ value is chosen sufficiently high.

2.1.4   Dynamic Behavior

When diffusion–reaction models of catalyst pellets contain nonlinearities due to nonlinear reaction kinetics or thermal effects, their solutions may exhibit unusual behavior, such as multiplicity of solutions for steady state models, and self-sustained oscillations for dynamic models (Aris, 1975). The case of pellets with Dirac-delta distributions is no exception to this. Brunovska (1987, 1988) and Barto et al. (1991) analyzed the steady-state and dynamic behavior of pellets with Dirac and step-type distributions where a single reaction

$$A_1 + \nu_2 A_2 \rightarrow \text{products} \tag{2.60}$$

following Langmuir–Hinshelwood kinetics

$$r = \frac{kK_1 K_2 C_1 C_2}{(1 + K_1 C_1 + K_2 C_2)^2} \tag{2.61}$$

occurs.

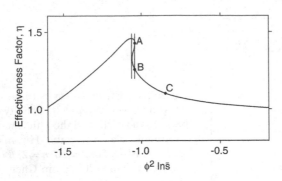

Figure 2.11. Effectiveness factor η as a function of the parameter $\phi^2 \ln \bar{s}$. Reaction $A_1 + \nu_2 A_2 \to$ products, following Langmuir–Hinshelwood kinetics according to equation (2.61), $n = 1, \sigma_1 = 1, \sigma_2 = 30, \nu_2 C_{f,1}/C_{f,2} = 10$, $D_{e,1}/D_{e,2} = 0.12$. (From Brunovska, 1987.)

For Dirac-delta catalysts of cylindrical shape, the dependence of the effectiveness factor on the parameter $\phi^2 \ln \bar{s}$ is shown in Figure 2.11. One stable steady state exists to the left of point A and to the right of point C. Between points B and C there is a region of unique unstable steady states surrounded by a limit cycle. Note that in the region between the two vertical lines the model exhibits three steady states, of which only that on the upper branch is stable. It should be noted that contrary to the steady-state solution, the transient solution depends independently on the Thiele modulus ϕ and the catalyst location $\bar{s}$. Therefore the stability character of the steady states in Figure 2.11 also changes with both ϕ and $\bar{s}$. In particular, it was found that the oscillatory region BC shrinks as the catalyst location $\bar{s}$ approaches the surface of the pellet.

Specifically, as the catalyst location moves towards the pellet external surface, the self-sustained oscillations first decrease in amplitude and then disappear with the formation of a stable steady state. The bifurcation diagram for Dirac-delta catalysts forms an envelope contained within the solid lines in Figure 2.12. The lower branch corresponds to point B and the upper branch to point C of Figure 2.11. Within the envelope periodic oscillations were observed, while stable steady states were obtained in the region outside the envelope. As noted above, as the catalyst moves towards the external surface, the two branches merge, thus indicating that oscillations are not possible.

Barto et al. (1991) further investigated this model using a different numerical technique and revealed more complex dynamic behavior. The effect of deactivation on the dynamics of Dirac pellets was also investigated. It was found that the decrease of catalyst activity during deactivation has similar consequences to the decrease of Thiele modulus for nondeactivating catalysts (Brunovska, 1987). Starting from a point on the left of A (Figure 2.11) where only one stable steady state exists, due to deactivation the region of periodic oscillations is reached. As deactivation continues, oscillations die out and a region of unique stable steady states is attained again, to the right of point C.

For step-type catalysts, the bifurcation diagram is somewhat deformed as compared to the Dirac catalyst (dashed lines in Figure 2.12). For a fixed catalyst location, the values of the lower and upper limits of the Thiele modulus within which periodic oscillations occur both *increase* from Dirac to step catalyst. It was also found that while for a Dirac-delta distribution the value of the Thiele modulus

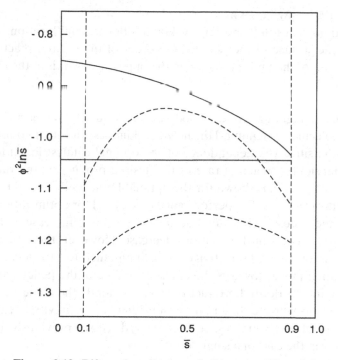

Figure 2.12. Bifurcation diagram. Solid curve, Dirac-delta activity distribution; dashed curve, step-type distribution. Reaction $A_1 + v_2 A_2 \rightarrow$ products follows Langmuir–Hinshelwood kinetics according to equation (2.61), $n = 1$, $\Delta = 0.1$, $\sigma_1 = 1$, $\sigma_2 = 30$, $v_2 C_{f,1}/C_{f,2} = 10$, $D_{e,1}/D_{e,2} = 0.12$. Periodic oscillations are observed within the envelopes. (From Brunovska, 1988.)

does not affect the amplitude of oscillations significantly, for a step distribution the amplitude increases as the Thiele modulus decreases. Furthermore, while for the Dirac distribution the relation of steady-state effectiveness factor to $\phi^2 \ln \bar{s}$ is independent of catalyst location $\bar{s}$ (see Figure 2.11), for step-type catalysts it depends on both step location and width.

2.2 Multiple Reactions

2.2.1 Isothermal Conditions

When more than one reaction is involved, other catalyst performance indexes related to the production of desired species become important in addition to the effectiveness factor, viz. selectivity and yield. The case where two reactions, either consecutive or parallel, following arbitrary kinetics occur was treated by Vayenas and Pavlou (1987b) to determine the catalyst distribution which maximizes selectivity. In all cases, the optimal distribution is an appropriately located Dirac-delta function.

Parallel reactions: $A_1 \rightarrow A_2$, $A_1 \rightarrow A_3$

For parallel reactions that follow power-law kinetics the optimal location of the Dirac-delta is at the surface of the pellet if the order of the desired reaction is higher than the order of the undesired one; in the opposite case it is at the center of the pellet. Egg-white distributions can be optimal for more complex kinetics. As an example, in Figure 2.13 we consider the case where the desired reaction follows bimolecular Langmuir–Hinshelwood kinetics, while the undesired reaction follows unimolecular Langmuir–Hinshelwood kinetics, and the two involve different adsorption sites. The dependence of the optimal catalyst location, selectivity, and dimensionless reaction rate of the desired product on the ratio of Thiele moduli, $\kappa = (\phi_2/\phi_1)^2$, is shown for the optimal Dirac, uniform, and external shell catalyst distributions. The performance index used for optimization was selectivity. When the undesired reaction becomes faster (i.e., κ increases), the selectivities for all three types of distributions decrease. However, the Dirac-type catalyst always gives the highest selectivity value. Simultaneously, the location of the Dirac distribution moves towards the external surface of the pellet. The net rate of production for the desired product is also consistently the highest for the Dirac distribution. It is worth noting that the sensitivity of selectivity to catalyst location behaves qualitatively in the same manner as discussed previously for the effectiveness factor in the case of a single reaction.

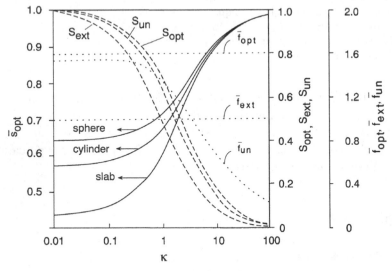

Figure 2.13. Effect of the ratio of Thiele moduli, κ, on the optimal catalyst location, $\bar{s}_{opt}$ (solid lines), global selectivity S (dashed lines), and dimensionless rate of production of the desired product A_2, $\bar{f}$ (dotted lines), for the optimal distribution (S_{opt}, $\bar{f}_{opt}$), uniform distribution for slab geometry (S_{un}, $\bar{f}_{un}$), and external shell distribution (S_{ext}, $\bar{f}_{ext}$). Parallel reactions: $A_1 \xrightarrow{1} A_2$, $A_1 \xrightarrow{2} A_3$; $f_1(u_1) = (1+\sigma_1)^2 u_1/(1+\sigma_1 u_1)^2$, $f_2(u_1) = (1+\sigma_2)u_1/(1+\sigma_2 u_1)$, $\sigma_1 = 5$, $\sigma_2 = 20$, $\phi_1[(n+1)/g_v]^{1/2} = 1$, $D_{e,1}/D_{e,2} = 1$. Optimization is based on selectivity. (From Vayenas and Pavlou, 1987b.)

Consecutive reactions: $A_1 \to A_2 \to A_3$

When two consecutive reactions occur, where the desired one follows bimolecular Langmuir–Hinshelwood kinetics and the undesired one follows linear kinetics, the Dirac catalyst distribution again gives the highest selectivity for all values of κ, as illustrated in Figure 2.14. The difference in selectivity obtained with different distributions (S_{opt}, S_{un}, and S_{ext}) increases with κ. The optimal distribution not only gives the highest selectivity, which is the optimization index, but also the highest net rate of production of the desired species. However, it should be noted that the optimal catalyst location for selectivity maximization is in general different from the optimal catalyst location for maximum production of the desired product. For the reaction system considered, the catalyst location is rather insensitive to changes in κ, except for sufficiently high values of κ, where the optimal location moves rapidly towards the center of the pellet, because diffusional resistances slow down the increasingly faster first-order undesired reaction. Although a Dirac-delta function is always the optimal distribution, the actual performance of Dirac catalysts will depend strongly on the type of kinetics that the reactions follow, and can be quite different than the specific cases discussed above.

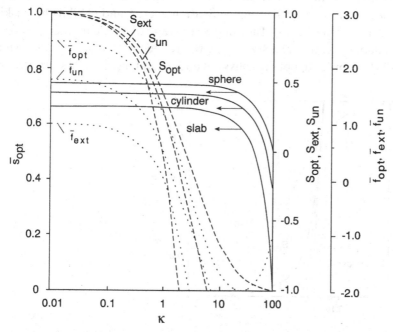

Figure 2.14. Effect of the ratio of Thiele moduli, κ, on the optimal catalyst location $\bar{s}_{opt}$ (solid lines), global selectivity, S (dashed lines), and dimensionless rate of production of the desired product A_2, $\bar{f}$ (dotted lines), for the optimal distribution (S_{opt}, $\bar{f}_{opt}$), uniform distribution for slab geometry (S_{un}, $\bar{f}_{un}$), and external shell distribution (S_{ext}, $\bar{f}_{ext}$). Consecutive reactions: $A_1 \xrightarrow{1} A_2 \xrightarrow{2} A_3$; $f_1(u_1) = (1+\sigma)^2 u_1/(1+\sigma u_1)^2$, $f_2(u_2) = u_2$, $\sigma = 8$, $\phi_1[(n+1)/g_v]^{1/2} = 1$, $D_{e,1}/D_{e,2} = 1$, $u_{f,2} = 1$. Optimization is based on selectivity. (From Vayenas and Pavlou, 1987b.)

2.2.2   Nonisothermal Conditions

The improved performance of Dirac catalysts, as demonstrated in previous sections, arises from the appropriate manipulation of concentration gradients which develop within the pellet. When detrimental, they are avoided by positioning the catalyst on the surface of the pellet. When they are beneficial, however, one can take advantage of them by placing the catalyst at an appropriate location inside the pellet. Temperature gradients are also manipulated in a similar way. However, their effect is generally more pronounced because of the exponential dependence of reaction rates on temperature. Similar to the isothermal case, parallel, consecutive, and also triangular reaction networks, occurring in nonisothermal pellets, show optimum performance when the catalyst is located at a specific position which depends upon the physicochemical parameters of the system (Morbidelli et al., 1984a; Vayenas and Pavlou, 1988; Vayenas et al., 1989; Pavlou and Vayenas, 1990a; Letkova et al., 1994). In this case, the activation energies and heats of reaction are also involved.

Parallel reactions: $A_1 \rightarrow A_2$, $A_1 \rightarrow A_3$
As an example consider two parallel reactions, where the desired one is exothermic, has low activation energy, and exhibits second-order kinetics, while the undesired one is endothermic, has high activation energy, and exhibits linear kinetics (Vayenas and Pavlou, 1988). The effect of the ratio of Thiele moduli, $\kappa = (\phi_2/\phi_1)^2$, on the optimal location and selectivity, where the latter is optimized, is shown in Figure 2.15. In order to obtain a physical understanding of this system, it should be

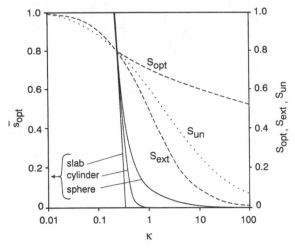

Figure 2.15. Effect of the ratio of Thiele moduli, κ, on the optimal catalyst location $\bar{s}_{opt}$ (solid lines), optimal global selectivity S_{opt}, and global selectivity obtained with uniform (S_{un}) and external shell (S_{ext}) catalyst distributions. Parallel reactions: $A_1 \xrightarrow{1} A_2$ (second order), $A_1 \xrightarrow{2} A_3$ (first order); desired product A_2; $\beta_1 = 0.1$, $\beta_2 = -0.7$, $\gamma_1 = 5$, $\gamma_2 = 25$, $\phi_1[(n+1)/g_v]^{1/2} = 1$. Optimization is based on selectivity. (From Vayenas and Pavlou, 1988.)

realized that increasing reactant concentration and decreasing temperature favor the desired reaction. For small κ values, the rate of the undesired reaction is relatively small and the desired exothermic reaction creates higher temperatures inside the pellet, if the catalyst is located there. This however has a detrimental effect on selectivity because of the relative activation energies, which adds to the detrimental concentration effect. Therefore, the optimum catalyst location is not the surface of the pellet. For large κ values, the undesired reaction is fast and overcomes the thermal effect of the exothermic desired reaction. As a result, if the catalyst is located inside the pellet, the temperature decreases, which benefits the desired reaction. This thermal effect easily surmounts the always negative concentration effect, and consequently the optimum catalyst location moves inside the pellet.

Consecutive reactions: $A_1 \rightarrow A_2 \rightarrow A_3$
The case of two consecutive exothermic reactions with linear kinetics and higher activation energy for the first desired reaction is illustrated in Figure 2.16 (Vayenas et al., 1989). Since the reaction orders are the same for both reactions, in this case it is clear that concentration gradients do not affect selectivity at all. Increased temperature favors the desired reaction, since its activation energy is larger than

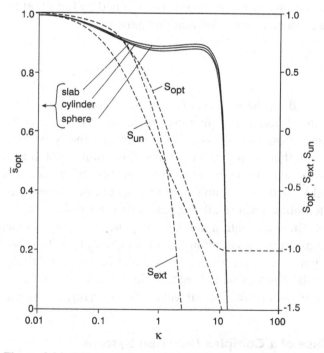

Figure 2.16. Effect of the ratio of Thiele moduli κ on the optimal catalyst location $\bar{s}_{opt}$ (solid lines), optimal global selectivity S_{opt}, and global selectivity obtained with uniform, (S_{un}) and external shell (S_{ext}) catalyst distributions. Consecutive reactions: $A_1 \xrightarrow{1} A_2$ (first order), $A_2 \xrightarrow{2} A_3$ (first order); desired product A_2; $\beta_1 = 0.4$, $\beta_2 = 0.1$, $\gamma_1 = 10$, $\gamma_2 = 5$, $\phi_1[(n+1)/g_v]^{1/2} = 1$, $D_{e,1}/D_{e,2} = 1$, $u_{f,2} = 1$. Optimization is based on selectivity. (From Vayenas et al., 1989.)

that of the undesired one, i.e. $\gamma_1 > \gamma_2$. Such temperatures are obtained by placing the catalyst inside the pellet. For high values of κ, which indicates that the undesired reaction is so fast that it approaches diffusion control, the optimal catalyst location moves rapidly towards the center of the pellet.

A system of multiple reactions of great industrial importance, the ethylene epoxidation reaction network, was investigated by Morbidelli et al. (1984a). Ethylene reacts over silver catalyst to give ethylene oxide, which is the desired product, and carbon dioxide, which is the undesired by-product. Ethylene oxide is also combusted, but this reaction is less important than the combustion of ethylene, and so it was not included in the above study. It was shown that the selectivity to ethylene oxide is maximized when the catalyst is located at the external surface of the pellet. This is expected for parallel exothermic reactions with similar kinetics when the undesired reaction has a higher activation energy than the desired one. The maximum selectivity is attained at the surface of the pellet, even if combustion of ethylene oxide is included (Pavlou and Vayenas, 1990a). However, if the net production rate of ethylene oxide is to be maximized, then subsurface locations are optimal, a fact which arises from the nonisothermality of the pellet.

Numerical optimization

A numerical optimization technique has been developed by Baratti et al. (1990), and applied to the consecutive–parallel reaction network

$$A_1 \rightarrow A_2 \rightarrow A_3$$
$$A_1 \rightarrow A_4. \tag{2.62}$$

The technique is based on the method of orthogonal collocation over finite elements, whereby the differential equations are reduced to a set of nonlinear algebraic equations, whose solution is obtained through the Newton–Raphson method. It was shown that the number of collocation points used in the calculations is critical, and if it is too small, then erroneous results can arise. In Figure 2.17, the numerically calculated optimal activity distribution is shown for the reacting system noted above, where all the reactions are exothermic and follow first-order kinetics. The optimization index is the pellet effectiveness factor, and it corresponds to maximum consumption rate of component A_1. The shape of the obtained distribution closely resembles a Dirac delta function, in full agreement with analytical calculations. This numerical technique will also be used in the optimization of complex reacting systems, and will be discussed further in section 2.5.4.

2.3   The General Case of a Complex Reaction System

Optimization methods based on the analysis of the performance-index integral [e.g. equation (2.11)] and of the reaction rate function $f(u)$ cannot be readily extended to complex reaction systems, which are frequently encountered in industrial applications. For this purpose, a method has been developed based on a necessary condition for optimality using variational techniques (Wu et al., 1990a;

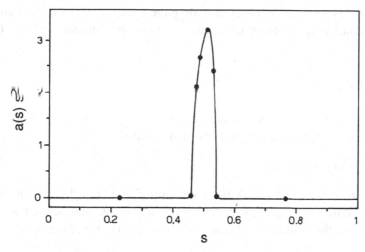

Figure 2.17. Optimal catalyst activity distribution for effectiveness factor maximization. Consecutive–parallel first-order reactions: $\phi_1 = 0.5$, $\phi_2 = 1$, $\phi_3 = 0.5$, $\beta_1 = 0.2$, $\beta_2 = 0.12$, $\beta_3 = 0.18$, $\gamma_1 = 20$, $\gamma_2 = 10$, $\gamma_3 = 20$, $D_{e,1}/D_{e,2} = 1$, $u_{f,2} = 0$. (From Baratti et al., 1990.)

Morbidelli et al., 1991). Before proceeding to analyze complex reaction systems, it is first illustrated for the simple case of a single isothermal reaction.

2.3.1 An Illustrative Example

Definition of optimization problem
As described in section 2.1.1, the relevant dimensionless equations for a single isothermal reaction with no external mass transfer resistance are as follows. The reactant mass balance gives

$$L[u] = \frac{1}{s^n}\frac{d}{ds}\left(s^n\frac{du}{ds}\right) = \phi^2 a(s) f(u) \tag{2.63}$$

along with BCs

$$s = 0: \quad \frac{du}{ds} = 0 \tag{2.64a}$$

$$s = 1: \quad u = 1. \tag{2.64b}$$

From the definition of the activity distribution function,

$$a(s) = k(s)/\bar{k}, \tag{2.65}$$

while the constraint on $a(s)$ arising from its definition requires

$$(n+1)\int_0^1 a(s)s^n\, ds = 1. \tag{2.66}$$

The effectiveness factor is given by

$$\eta = (n+1)\int_0^1 f(u)a(s)s^n\, ds = \frac{n+1}{\phi^2}\left(\frac{du}{ds}\right)_{s=1}. \tag{2.67}$$

The optimization problem consists in finding the function $a(s)$ which maximizes the performance index η under the constraint given by equations (2.63), (2.64), and (2.66).

The necessary condition for optimality

Let us assume that $a^*(s)$ is the optimal catalyst distribution which maximizes the objective functional (2.67). Any other distribution $a(s)$ in the neighborhood of $a^*(s)$ can then be expressed as a perturbation about $a^*(s)$:

$$a(s) = a^*(s) + \delta a \tag{2.68}$$

where both $a(s)$ and $a^*(s)$ satisfy the constraint (2.66). The dimensionless concentration profile $u(s)$, resulting from $a(s)$, can also be represented as a perturbation about the optimal concentration profile $u^*(s)$ resulting from $a^*(s)$ as follows:

$$u(s) = u^*(s) + \delta u. \tag{2.69}$$

If the perturbation is chosen sufficiently small, then a first-order expansion about a^* is sufficient, and linearization of equation (2.63) about a^* yields

$$L[\delta u] = \phi^2(a^* f'^* \delta u + f^* \delta a) \tag{2.70}$$

with the BCs

$$s = 0: \qquad \frac{d\delta u}{ds} = 0 \tag{2.71a}$$

$$s = 1: \qquad \delta u = 0 \tag{2.71b}$$

where

$$f^* = f(u^*) \quad \text{and} \quad f'^* = \left.\frac{df(u)}{du}\right|_{u=u^*}. \tag{2.72}$$

Similarly, for the effectiveness factor,

$$\delta\eta = \eta[a(s)] - \eta[a^*(s)] = (n+1)\int_0^1 (a^* f'^* \delta u. + f^*\delta a)s^n \, ds. \tag{2.73}$$

In order to obtain the complete influence of the perturbation δa on the objective functional, we need to adjoin the linearized constraint (2.70) to the objective functional by introducing a Lagrange multiplier $\lambda(s)$. This is done by adding the identity

$$(n+1)\int_0^1 \lambda\{L[\delta u] - \phi^2(a^* f'^* \delta u + f^*\delta a)\}s^n \, ds = 0 \tag{2.74}$$

to equation (2.73), to get

$$\delta\eta = (n+1)\int_0^1 \{(1-\phi^2\lambda)f^*\delta a + (1-\phi^2\lambda)a^* f'^* \delta u + \lambda L[\delta u]\}s^n \, ds. \tag{2.75}$$

Since the Lagrange multiplier is arbitrary, we can define it as follows:

$$L[\lambda] = (\phi^2\lambda - 1)a^* f'^* \tag{2.76}$$

with BCs

$$s = 0, \qquad \frac{d\lambda}{ds} = 0 \tag{2.77a}$$

$$s = 1: \qquad \lambda = 0. \tag{2.77b}$$

The last term on the right-hand side of equation (2.75), when integrated twice by parts, becomes

$$\int_0^1 \lambda L[\delta u]s^n \, ds = \int_0^1 \delta u \, L[\lambda]s^n \, ds \tag{2.78}$$

Substituting the above relation into equation (2.75) and using equations (2.76) and (2.77) leads to

$$\delta\eta = (n+1)\int_0^1 G^* \delta a \, s^n \, ds \tag{2.79}$$

where

$$G^* = (1 - \phi^2\lambda) f^* \tag{2.80}$$

will be referred to subsequently as the G function. The asterisk indicates that all quantities in the G function are evaluated at the assumed optimal activity distribution, $a^*(s)$.

Equation (2.79) indicates the direct influence of δa on $\delta\eta$. The necessary condition for optimality is that $\delta\eta \le 0$ for all possible small variations δa. Using equation (2.68), this becomes

$$\int_0^1 G^* a^*(s)s^n \, ds \ge \int_0^1 G^* a(s)s^n \, ds. \tag{2.81}$$

Shape of the optimal catalyst distribution
Having assumed that $a^*(s)$ is the optimal distribution, we prove that $a^*(s)$ can satisfy the necessary condition (2.81) for optimality only if it is substantially nonzero solely at the point $\bar{s}$ where the corresponding $G^*(s)$ attains its maximum value over $0 < s < 1$. By *substantially nonzero* we mean that

$$\int_{\bar{s}-\varepsilon}^{\bar{s}+\varepsilon} a^*(s)s^n \, ds \ne 0 \tag{2.82}$$

for arbitrary small $\varepsilon > 0$. The above statement is proven by contradiction, i.e. by showing that if it is not true, and $a^*(s)$ is substantially nonzero at least at one point $s' \ne \bar{s}$ where

$$G^*(s') < G^*(\bar{s}) \tag{2.83}$$

then the catalyst distribution $a^*(s)$ cannot be optimal.

For this purpose, let us define a new distribution $a(s)$, which is a small perturbation of $a^*(s)$, satisfying the constraint (2.66):

$$a(s) = \begin{cases} a^*(s) + \gamma[s + s' - \bar{s}]^n a^*(s + s' - \bar{s})/s^n & \text{for} \quad \bar{s} - \varepsilon < s < \bar{s} + \varepsilon \\ (1 - \gamma)a^*(s) & \text{for} \quad s' - \varepsilon < s < s' + \varepsilon \\ a^*(s) & \text{otherwise.} \end{cases}$$

(2.84)

Since $\gamma > 0$ is arbitrarily small, $a(s)$ can be made infinitesimally close to $a^*(s)$, as required by equation (2.68). Therefore the necessary condition for optimality, (2.81), can be used. From equation (2.84) it follows that

$$\int_0^1 G^*(s)a(s)s^n \, ds = \gamma \int_{\bar{s}-\varepsilon}^{\bar{s}+\varepsilon} [s + s' - \bar{s}]^n a^*(s + s' - \bar{s})G^*(s) \, ds$$

$$+ \int_{(0,1)\backslash(s'-\varepsilon,s'+\varepsilon)} a^*(s)G^*(s)s^n \, ds$$

$$+ (1 - \gamma) \int_{s'-\varepsilon}^{s'+\varepsilon} a^*(s)G^*(s)s^n \, ds.$$

(2.85)

From equation (2.83) and the continuity of $G^*(s)$, it follows that for sufficiently small $\varepsilon > 0$

$$G^*(s - s' + \bar{s}) > G^*(s) \qquad \text{for} \quad s' - \varepsilon < s < s' + \varepsilon.$$

(2.86)

Using this equation, we determine a lower bound on the first integral on the right-hand side of equation (2.85). For this, it is convenient to introduce the new variable $z = s + s' - \bar{s}$, and then from equation (2.86) we obtain

$$\int_{\bar{s}-\varepsilon}^{\bar{s}+\varepsilon} [s + s' - \bar{s}]^n a^*(s + s' - \bar{s})G^*(s) \, ds > \int_{s'-\varepsilon}^{s'+\varepsilon} G^*(s)a^*(s)s^n \, ds.$$

(2.87)

Using equation (2.87), equation (2.85) yields

$$\int_0^1 G^*(s)a(s)s^n \, ds > \int_0^1 G^*(s)a^*(s)s^n \, ds$$

(2.88)

which violates the necessary condition (2.81) for optimality. Since the distribution $a(s)$, which is a small perturbation about $a^*(s)$, violates (2.81), $a^*(s)$ cannot be optimal. Therefore, in order for $a^*(s)$ to be optimal, it has to be substantially nonzero *only* at the point $\bar{s}$ where the corresponding $G^*(s)$ is maximum, which means that $a^*(s)$ must be a *Dirac-delta function*. Although it is highly unlikely that $G^*(s)$ will attain its maximum simultaneously at more than one point, it should be mentioned that in this case the optimal distribution would take the form of multiple Dirac-delta functions (Pavlou et al., 1991; Morbidelli et al., 1991; see also section 2.5.5).

It is worth noting that the above proof fails when the optimal activity distribution $a^*(s)$ generates a function $G^*(s)$ which is *constant* and *equal to its maximum* value G_m^* for all values of s where $a^*(s)$ is nonzero:

$$G^*(s) = G_m^* \quad \text{for} \quad a^*(s) > 0$$
$$G^*(s) \le G_m^* \quad \text{for} \quad a^*(s) = 0. \tag{2.89}$$

In this case it is not possible to devise a distribution $a(s)$, infinitesimally close to $a^*(s)$, which violates the necessary condition (2.01). It is in fact impossible to move an infinitesimally small portion of $a^*(s)$ from an interval $(s' - \varepsilon, s' + \varepsilon)$ to another $(s' - \varepsilon, s' + \varepsilon)$ where $G^*(s)$ is larger, as we did in equation (2.84). Therefore, in this particular case, the optimal distribution does not need to exhibit the shape of a Dirac delta function. Although this situation is rather unlikely in practice, at least one example exists (Keller et al., 1984), which will be discussed in section 5.3, since it refers to a catalytic membrane.

2.3.2 General Reaction System

The above procedure is now extended to a general reaction system comprising an arbitrary number of reactions with arbitrary kinetics, occurring in a nonisothermal pellet with finite external mass and heat transfer resistances.

Definition of optimization problem

Consider a general reaction scheme where J independent reactions occur and I components $(A_1, A_2, \ldots, A_I)$ are involved:

$$\sum_{i=1}^{I} v_{ij} A_i = 0 \quad (j = 1, \ldots, J).$$

Each reaction follows arbitrary kinetics, hence

$$r_j = r_j(C_1, C_2, \ldots, C_I, T) \equiv r_j(\mathbf{c}, T), \tag{2.90}$$

where $\mathbf{c}$ is the vector of concentrations. The steady-state mass and energy balances in a porous catalyst pellet, allowing for external transport resistances, are given by

$$D_{e,i} \frac{1}{x^n} \frac{d}{dx} \left(x^n \frac{dC_i}{dx} \right) = a(x) \sum_{j=1}^{J} v_{ij} r_j(\mathbf{c}, T) \quad (i = 1, \ldots, I) \tag{2.91}$$

$$\lambda_e \frac{1}{x^n} \frac{d}{dx} \left(x^n \frac{dT}{dx} \right) = -a(x) \sum_{j=1}^{J} (-\Delta H_j) r_j(\mathbf{c}, T) \tag{2.92}$$

along with the BCs

$$x = 0: \quad \frac{dC_i}{dx} = 0 \quad (i = 1, \ldots, I) \tag{2.93a}$$

$$x = R: \quad D_{e,i} \frac{dC_i}{dx} = k_{g,i}(C_{f,i} - C_i) \quad (i = 1, \ldots, I) \tag{2.93b}$$

$$x = 0: \quad \frac{dT}{dx} = 0 \tag{2.93c}$$

$$x = R: \quad \lambda_e \frac{dT}{dx} = h(T_f - T). \tag{2.93d}$$

In dimensionless form, the above equations become

$$L[u_i] = \frac{1}{s^n}\frac{d}{ds}\left(s^n\frac{du_i}{ds}\right) = a(s)D_i\sum_{j=1}^{J}v_{ij}\phi_j^2 f_j(\mathbf{u},\theta) \qquad (i=1,\ldots,I) \quad (2.94)$$

$$L[\theta] = \frac{1}{s^n}\frac{d}{ds}\left[s^n\frac{d\theta}{ds}\right] = -a(s)\sum_{j=1}^{J}\beta_j\phi_j^2 f_j(\mathbf{u},\theta) \qquad\qquad (2.95)$$

$$s=0: \qquad \frac{du_i}{ds}=0 \qquad\qquad (i=1,\ldots,I) \qquad\qquad (2.96a)$$

$$s=1: \qquad \frac{du_i}{ds}=\text{Bi}_{m,i}(u_{f,i}-u_i) \quad (i=1,\ldots,I) \qquad\qquad (2.96b)$$

$$s=0: \qquad \frac{d\theta}{ds}=0 \qquad\qquad\qquad (2.96c)$$

$$s=1: \qquad \frac{d\theta}{ds}=\text{Bi}_h(1-\theta) \qquad\qquad\qquad (2.96d)$$

where

$$u_i = C_i/C_{f,1}, \qquad s=x/R, \qquad D_i = D_{e,1}/D_{e,i}$$

$$\phi_j^2 = r_j(C_{f,1}, C_{f,1}, \ldots, C_{f,1}, T_f)R^2/D_{e,1}C_{f,1}$$

$$f_j(\mathbf{u},\theta) = r_j(C_1, C_2, \ldots, C_I, T)/r_j(C_{f,1}, C_{f,1}, \ldots, C_{f,1}, T_f) \qquad (2.97)$$

$$\theta = T/T_f, \qquad \beta_j = (-\Delta H_j)D_{e,1}C_{f,1}/\lambda_e T_f$$

$$\text{Bi}_{m,i} = k_{g,i}R/D_{e,i}, \qquad \text{Bi}_h = hR/\lambda_e$$

and the activity distribution $a(s)$ satisfies constraint (2.66). Several indexes can be used to evaluate the performance of a catalyst pellet when multiple reactions occur. These include the effectiveness factor, which represents the overall consumption rate of a certain reactant; the selectivity, which represents the fraction of a given reactant to a desired product; and the yield, which represents the overall production rate of a certain product.

The effectiveness factor with respect to the Mth component is defined as the ratio between the actual consumption rate of the component and the rate which would have existed in the absence of *all* transport resistances:

$$\eta_M = \frac{(n+1)\int_0^1 a(s)\left[\sum_{j=1}^{J}v_{Mj}\phi_j^2 f_j(\mathbf{u},\theta)\right]s^n\,ds}{\sum_{j=1}^{J}v_{Mj}\phi_j^2}. \qquad (2.98)$$

The selectivity of the Mth component (reactant) towards the Nth component (product) is defined as the ratio between the rate of production of component N and the rate of consumption of component M:

$$S_{NM} = \frac{\int_0^1 a(s)\left[\sum_{j=1}^{J}v_{Nj}\phi_j^2 f_j(\mathbf{u},\theta)\right]s^n\,ds}{\int_0^1 a(s)\left[\sum_{j=1}^{J}v_{Mj}\phi_j^2 f_j(\mathbf{u},\theta)\right]s^n\,ds}. \qquad (2.99)$$

The yield of desired product N with respect to reactant M is defined as the ratio between the actual rate of production of component N and the rate of consumption of reactant M which would have prevailed in the absence of all transport resistances:

$$Y_{NM} = \eta_M S_{NM} = \frac{(n+1)\int_0^1 a(s)\left[\sum_{j=1}^J \nu_{Nj}\phi_j^2 f_j\right]s^n ds}{\sum_{j=1}^J \nu_{Mj}\phi_j^2} \tag{2.100}$$

In order to simplify the notation, let us introduce the $(I+1)$-dimensional vectors

$$\mathbf{x} = \begin{bmatrix} u_1 \\ u_2 \\ \vdots \\ u_I \\ \theta \end{bmatrix}, \qquad \mathbf{w}(\mathbf{x}) = \begin{bmatrix} D_1 \sum_{j=1}^J \nu_{1j}\phi_j^2 f_j(\mathbf{x}) \\ D_2 \sum_{j=1}^J \nu_{2j}\phi_j^2 f_j(\mathbf{x}) \\ \vdots \\ D_I \sum_{j=1}^J \nu_{Ij}\phi_j^2 f_j(\mathbf{x}) \\ -\sum_{j=1}^J \beta_j \phi_j^2 f_j(\mathbf{x}) \end{bmatrix}, \qquad \mathbf{x}_f = \begin{bmatrix} 1 \\ u_{f,2} \\ \vdots \\ u_{f,I} \\ 1 \end{bmatrix} \tag{2.101}$$

and the diagonal matrix

$$\mathbf{A} = \begin{bmatrix} \mathrm{Bi}_{m,1} & 0 & & & \\ 0 & \mathrm{Bi}_{m,2} & 0 & & \\ & \ddots & \ddots & \ddots & \\ & & 0 & \mathrm{Bi}_{m,I} & 0 \\ & & & 0 & \mathrm{Bi}_h \end{bmatrix}. \tag{2.102}$$

The basic equations (2.94)–(2.96) can now be written concisely as follows:

$$L[\mathbf{x}] = a(s)\mathbf{w}(\mathbf{x}) \tag{2.103}$$

with BCs

$$s = 0: \qquad \frac{d\mathbf{x}}{ds} = \mathbf{0} \tag{2.104a}$$

$$s = 1: \qquad \frac{d\mathbf{x}}{ds} = \mathbf{A}(\mathbf{x}_f - \mathbf{x}). \tag{2.104b}$$

The expressions for the performance indexes [equations (2.98)–(2.100)] take the form

$$\eta_M = \frac{(n+1)\int_0^1 a(s)w_M(\mathbf{x})s^n ds}{D_M \sum_{j=1}^J \nu_{Mj}\phi_j^2} \tag{2.105}$$

$$S_{NM} = \frac{D_M \int_0^1 a(s)w_N(\mathbf{x})s^n ds}{D_N \int_0^1 a(s)w_M(\mathbf{x})s^n ds} \tag{2.106}$$

$$Y_{NM} = \eta_M S_{NM} = \frac{(n+1)\int_0^1 a(s)w_N(\mathbf{x})s^n ds}{D_N \sum_{j=1}^J \nu_{Mj}\phi_j^2}. \tag{2.107}$$

Since the expressions for the effectiveness factor (2.105) and yield (2.107) have the same form, the objective function in these cases, up to a constant multiplier, can be written in general as

$$U_1 = (n+1) \int_0^1 a(s) w_i(\mathbf{x}) s^n \, ds. \tag{2.108}$$

Similarly, for the selectivity the objective function is given by

$$U_2 = \frac{\int_0^1 a(s) w_i(\mathbf{x}) s^n \, ds}{\int_0^1 a(s) w_j(\mathbf{x}) s^n \, ds}. \tag{2.109}$$

The optimality conditions

The optimality conditions are obtained as in section 2.3.1 by linearizing the constraints (mass and energy balances) and the objective functional and adjoining them with an $(I+1)$-dimensional Lagrange multiplier vector. The G functions (G_1 for effectiveness factor and yield, and G_2 for selectivity) obtained are (Wu et al., 1990a)

$$G_1^* = w_i - \boldsymbol{\lambda}^T \mathbf{w} \tag{2.110}$$

$$G_2^* = \frac{\Omega_j w_i - \Omega_i w_j - \boldsymbol{\lambda}^T \mathbf{w}}{\Omega_j^2} \tag{2.111}$$

where

$$\Omega_k = (n+1) \int_0^1 a^*(s) w_k s^n \, ds. \tag{2.112}$$

The necessary conditions for optimality of effectiveness factor and yield are

$$\int_0^1 G_1^* a^*(s) s^n \, ds \geq \int_0^1 G_1^* a(s) s^n \, ds \tag{2.113}$$

and for selectivity

$$\int_0^1 G_2^* a^*(s) s^n \, ds \geq \int_0^1 G_2^* a(s) s^n \, ds. \tag{2.114}$$

We may observe that, even though the expressions for the G function are different for the two optimization problems, the *form* of the optimality condition is the same. The method utilized in the case of a single isothermal reaction to prove that the optimal catalyst distribution is a Dirac-delta function is based on the existence and form of the G function. Since the optimality conditions (2.113) and (2.114) have the same form as (2.81), and furthermore, no matter how complicated the G functions may be, when a^* is given, G_1^* and G_2^* are functions only of s, the same arguments can be followed also in this case.

Thus, it can be concluded that, *for any catalyst performance index (i.e. effectiveness, selectivity, or yield) and for the most general case of an arbitrary number of reactions, following arbitrary kinetics, occurring in a nonisothermal pellet, with finite external mass and heat transfer resistances, the optimal catalyst activity distribution is a Dirac-delta function.*

An alternative method to prove that the optimal distribution is a Dirac-delta function has been reported by Ye and Yuan (1992).

Evaluation of optimal catalyst location

Having established that the optimal distribution is a Dirac-delta function, it is now only required to determine its location. This can be conveniently done by seeking directly the Dirac-delta location which maximizes the desired performance index, as described below.

For a Dirac catalyst distribution

$$a(s) = \frac{\delta(s - \bar{s})}{(n+1)\bar{s}^n} \tag{2.115}$$

the effectiveness factor with respect to the Mth component (2.98) reduces to the *algebraic* equation

$$\eta_M = \frac{\sum_{j=1}^{J} \nu_{Mj} \phi_j^2 f_j(\bar{\mathbf{u}}, \bar{\theta})}{\sum_{j=1}^{J} \nu_{Mj} \phi_j^2}, \tag{2.116}$$

where $\bar{\mathbf{u}}$ and $\bar{\theta}$ represent the dimensionless species concentrations and temperature at the Dirac location $\bar{s}$. Similarly, the selectivity (2.99) of reactant M towards product N reduces to

$$S_{NM} = \frac{\sum_{j=1}^{J} \nu_{Nj} \phi_j^2 f_j(\bar{\mathbf{u}}, \bar{\theta})}{\sum_{j=1}^{J} \nu_{Mj} \phi_j^2 f_j(\bar{\mathbf{u}}, \bar{\theta})}, \tag{2.117}$$

and the yield (2.100) towards the desired product N with respect to reactant M to

$$Y_{NM} = \eta_M S_{NM} = \frac{\sum_{j=1}^{J} \nu_{Nj} \phi_j^2 f_j(\bar{\mathbf{u}}, \bar{\theta})}{\sum_{j=1}^{J} \nu_{Mj} \phi_j^2}. \tag{2.118}$$

In addition, the diffusion–reaction equations (2.94)–(2.96) reduce to a set of algebraic equations

$$\bar{u}_i - u_{f,i} + \left(\frac{1}{\text{Bi}_{m,i}} - \psi_n(\bar{s}) \right) \frac{D_i \sum_{j=1}^{J} \nu_{ij} \phi_j^2 f_j(\bar{\mathbf{u}}, \bar{\theta})}{n+1} = 0 \tag{2.119}$$

$$\bar{\theta} - 1 - \left(\frac{1}{\text{Bi}_h} - \psi_n(\bar{s}) \right) \frac{\sum_{j=1}^{J} \beta_j \phi_j^2 f_j(\bar{\mathbf{u}}, \bar{\theta})}{n+1} = 0 \tag{2.120}$$

where

$$\psi_n(\bar{s}) = \begin{cases} \bar{s} - 1 & \text{for} \quad n = 0 \\ \ln \bar{s} & \text{for} \quad n = 1 \\ 1 - 1/\bar{s} & \text{for} \quad n = 2. \end{cases} \tag{2.121}$$

For a given catalyst location $\bar{s}$, equations (2.119) and (2.120) are solved for $\bar{\mathbf{u}}$ and $\bar{\theta}$ and then the performance index of interest is calculated from equations (2.116)–(2.118). Caution should be exercised during the numerical solution, because the nonlinearity of the equations implies the possibility of multiple solutions. The optimal catalyst location $\bar{s}_{opt}$ is obtained by identifying the $\bar{s}$ value in the interval [0, 1] which maximizes the desired performance index. Note that for a given reacting system, the optimal catalyst location $\bar{s}_{opt}$ depends strongly on whether the performance index considered is effectiveness, selectivity, or yield.

2.4   Catalyst Dispersion Considerations

In sections 2.1–2.3, we have been concerned with optimization of the catalyst *activity* distribution. However, when preparing a catalyst we have control over the catalyst *loading* distribution. Therefore it is desirable to use catalyst loading directly as the optimization variable. For this, a relationship between catalyst activity and loading is needed. For structure-insensitive reactions, the activity of a catalyst is proportional to its surface area. Hence, it suffices to know the dependence of its surface area on its loading. At sufficiently low catalyst concentrations, the surface area of an active element is proportional to its loading, resulting in a linear relation between catalyst activity and loading. However, nonlinear dependence of surface area on loading can exist, especially at high catalyst concentrations. In this case, the optimization of the catalyst activity distribution should take this dependence into account. For structure-sensitive reactions, the relation between reaction rate and surface area depends strongly on the catalytic system, so that the optimal distribution problem has to be dealt with case by case.

2.4.1   Factors Affecting Catalyst Dispersion

The catalytic surface area of a supported catalyst (or equivalently its dispersion) depends significantly on its preparation procedure. A common technique for this is impregnation of the support with a precursor solution. The conditions of the impregnation step, as well as the subsequent steps of drying, calcination, and reduction, affect dispersion and hence the dependence of catalytic surface area on catalyst loading. More specifically, parameters that influence catalyst dispersion include the nature and concentration of precursor and support, the duration, temperature, and atmosphere (in contact with the catalyst) of the preparation steps, etc. (Delmon and Houalla, 1979). The effects of these parameters on catalyst properties are complex, and have been the topic of many studies over the years (cf. Schwarz et al., 1995). In the following, only a few representative cases are discussed, in order to demonstrate underlying phenomena affecting the active surface area of a supported catalyst.

Impregnation
During impregnation, the active agent is introduced into the support as a solution of a precursor compound, which is typically not its final form. Catalyst dispersion is affected in this step by the interaction of precursor and support (e.g. strong or weak adsorption). This in turn depends on the precursor properties and concentration, nature of the support, solution pH, ionic strength, and presence

of added or extraneous ions. The effects of these parameters on adsorption are discussed in more detail in Chapter 7. Generally, when strong interaction exists during impregnation, the precursor establishes a physical or chemical bond with the support surface. These interactions result in a near-atomic distribution of the precursor on the support, and thus in high dispersion.

When only weak precursor–support interaction exists, the support acts merely as a physical surface. In this case, the catalyst dispersion is typically lower than in the case of strong interaction. Metal particles develop through crystallization of precursor in the pore-filling solution during solvent evaporation. Crystallite sizes are dictated by mass transfer during precipitation–crystallization of the dissolved component, which in turn is controlled by the conditions during solvent evaporation (Le Page, 1987). Weak precursor–support interaction can be encountered in the following cases: (a) the support is inert, (b) the precursor ion has the same charge as the support surface, (c) the support is in a range of pH where it is not charged, (d) the adsorption sites of the support are occupied.

The surface of α-Al_2O_3 is inert because it is dehydroxylated and hence cannot develop surface charge. Hence, catalysts based on α-Al_2O_3 have typically low dispersion (cf. Gavriilidis et al., 1993b). Interaction between support and catalyst precursor can be affected by the combination of precursor and support used. When the support surface is negatively charged in solution, as is the case with silica (Brunelle, 1978), the metal will not be adsorbed if it is contained in the anion, as in H_2PtCl_6, but will be adsorbed if it is present in the cation, as in $[Pt(NH_3)_4]Cl_2$. No adsorption occurs even when $[Pt(NH_3)_4]Cl_2$ is used as the precursor if the pH is below 6 (Benesi et al., 1968). Silica has a zeta-potential value of almost zero for pH 0–6, indicating that in this pH range the charge of the surface is practically zero, and hence it behaves as an inert support (Anderson, 1975; Brunelle 1978). The above picture of the support–precursor interaction is simplistic, but it provides a framework within which the behavior of a large number of support–precursor systems can be understood.

Finally, every support material has a finite adsorption capacity. If the amount of metal ion to be deposited is larger than the adsorption capacity at a given pH, then the excess amount will deposit without any interaction with the support surface (Che and Bonneviot, 1988). For example, for iridium–titania systems, it has been observed that two types of particles are formed: small particles produced from precursor ion-exchanged on the support surface, and large particles produced by precursor not interacting with the support surface (Van Tiep et al., 1986).

Drying

The drying process can affect metal particle size, primarily when weak support–precursor interaction exists. In this case, catalyst particles are formed by crystallization of the precursor from the pore-filling liquid during solvent evaporation. The ratio of nucleation rate to crystal growth rate determines the crystal size; if it is large, the crystal size is small (Le Page, 1987). A high internal surface area favors heterogeneous nucleation. The nucleation–crystallization process is influenced by mass and heat transfer in the porous structure. When evaporation of the solvent is fast compared to solute diffusion, the precursor concentration increases at the air–liquid interface, which gradually recedes to the pellet interior. Hence, particles deposit at the interface, leading to relatively homogeneous deposition and particle size throughout the pellet.

If the evaporation is slow, the particle size depends on the pore structure as well as on the precursor concentration of the impregnating solution. Solvent removal concentrates the solution to the point where precursor crystallization begins, providing nuclei in the pores still containing solution. The crystallites will grow as long as there is precursor-containing solution surrounding them. Large pores can accommodate large solution volume, and hence crystallite size is related to pore size. Indeed, it has been observed that supports with large pore size give rise to large catalyst particles (Dorling et al., 1971; Hamada et al., 1987). In addition, the pore structure can affect the availability of precursor for crystallite growth, by leading to formation of isolated clusters of liquid-filled pores.

Similar arguments can be used to explain the effect of precursor concentration on particle size (Dorling and Moss, 1967; Dorling et al., 1971; Moss et al., 1979). If the impregnating solution is dilute and solute diffusion is fast, many pores empty before crystallization begins. Increasing the concentration of the impregnating solution increases the number of nuclei, and at a certain precursor concentration, crystallization starts in practically all pores. Up to this point, the mean crystallite size is almost constant, and the catalyst surface area is proportional to the catalyst loading. At higher precursor concentration, though, the number of crystallites remains constant but the crystallite size increases, and hence the catalyst surface area does not increase linearly with further catalyst loading.

Calcination

The calcination step, which is usually carried out in oxidizing atmosphere, can lead to a number of transformations (Che and Bennett, 1989): precursor decomposition and formation of an oxide species, bonding of the formed oxide with the support, removal of some of the elements introduced during the preparation, decomposition of the precursor ionic complex, ligand exchange reactions between surface anchoring groups and ligands bound to the metal ion, elimination of carbonaceous impurities, and sintering of the precursor compound or the formed oxide species.

For platinum–silica catalysts, crystallite size was found to increase with calcination temperature (Dorling et al., 1971; Brunelle et al., 1976). When catalysts were obtained by chloroplatinic acid impregnation (no support–precursor interaction), the increase was more substantial than with catalysts prepared by amine complex (interaction through ion exchange). Furthermore, dispersion loss started at lower temperature for the former type of catalysts. For Pt/γ-Al_2O_3 catalysts, Bournonville et al. (1983) found that the metal dispersion exhibited a maximum as a function of calcination temperature. It was proposed that increasing calcination temperature facilitated the formation of an oxochlorinated platinum complex, which starts to decompose above a certain temperature. The formation of this complex was crucial for obtaining high dispersion. This complex was strongly bound to the carrier and led to the formation of a well-dispersed metallic phase upon reduction. Catalysts that were reduced without previous calcination had significantly lower dispersion than those subjected to calcination before reduction, as also found by Carballo et al. (1978). Dispersion loss at high calcination temperature was more pronounced at higher platinum loadings.

High-temperature treatments such as calcination and reduction can lead to sintering of the precursor compound or the formed oxide species. This results in larger particles and loss of catalyst dispersion. Excellent reviews of sintering phenomena have appeared in the literature (Wanke and Flynn, 1975; Ruckenstein and Dadyburjor, 1983; Bartholomew, 1993). High-temperature treatments can also lead to increase of catalyst dispersion. It has been demonstrated that oxidation treatments or oxidation–reduction cycles at high temperature can result in redispersion of large metal particles (cf. Ruckenstein and Malhotra, 1976; Wang and Schmidt, 1981). This has been attributed to the lower surface tension of the metal oxide than of the metal, causing a flattened structure to be formed from a hemispherical particle of the reduced metal, which is readily reconstructed into smaller hemispherical particles upon reduction of the oxidized structures (Lee and Ruckenstein, 1983).

Reduction

Reduction in hydrogen is commonly employed for catalyst activation, during which the metal precursor compound or its oxide is transformed into the metallic state (Foger, 1984). The interaction of the oxide with the support, which depends on the phenomena that have taken place during the preceding calcination step, can also affect the reduction step (Delmon and Houalla, 1979). Various effects of reduction temperature on particle size have been observed.

Increasing the temperature of reduction was found to result in an increase in particle size for Pt/γ-Al_2O_3 catalysts prepared by chloroplatinic acid impregnation (Wilson and Hall, 1970). Only a small effect of reduction temperature on platinum particle size was observed for catalysts prepared by a platinum amine complex adsorption and reduced at 400–700°C (Brunelle et al., 1976). For Ni/Al_2O_3 catalysts, the nickel surface area showed a rather weak dependence on reduction temperature between 300 and 500°C, with a maximum occurring at about 350–400°C. Substantial loss of nickel surface area was obtained if the samples were calcined before reduction (Bartholomew and Farrauto, 1976). Other treatments before reduction can also affect catalyst dispersion. Sarkany and Gonzalez (1983) showed that initial pretreatment in helium results in Pt/γ-Al_2O_3 catalyst with dispersion considerably larger than that obtained after reduction. In this case, the dispersion reached a value of 60% at loading of 2 wt% and remained constant up to 6 wt% after helium pretreatment, whereas after reduction it decreased from 50% to 20% as the loading increased from 2 to 6 wt%.

2.4.2 Dependence of Catalytic Surface Area on Catalyst Loading

From the above discussion, it is clear that catalytic surface area depends strongly on the preparation conditions. A variety of experimental techniques are available to measure catalytic surface area (or equivalently catalyst dispersion) and catalyst loading (cf. Delannay, 1984). Such measurements have been performed for various catalysts such as Pt/SiO_2 (Dorling et al., 1971; Brunelle et al., 1976), Pt/Al_2O_3 (Basset et al., 1975), Rh/SiO_2 (Arakawa et al., 1984; Underwood and Bell, 1987), Rh/Al_2O_3 (Fuentes and Figueras, 1980), Pd/SiO_2 (Moss et al., 1979), Pd/Al_2O_3 (Scholten and Van Montfoort, 1962; Aben, 1968), Ag/SiO_2 (Seyed-monir et al., 1990), Ag/Al_2O_3 (Gavriilidis et al., 1993b), Ni/Al_2O_3 (Bartholomew et al., 1980; Huang and Schwarz, 1987), Co/Al_2O_3 (Reuel and Bartholomew, 1984), Ir/Al_2O_3 (Barbier and Marecot, 1981), Ru/Al_2O_3 (King, 1978), and V_2O_5/ZrO_2 (Chary et al., 1991).

A relationship between active element surface area per unit weight of catalyst pellet, A, and weight fraction of active catalyst, q, which can be used to represent many real situations is

$$A = \frac{pq}{1 + bq} \tag{2.122}$$

where p represents the specific surface area of active catalyst (square meters per gram of active catalyst) and p/b is surface area of active catalyst per unit weight of catalyst pellet at saturation, i.e. at high loadings. Note that from the active element surface area A, the catalyst dispersion D, defined as the fraction of active element atoms available on crystallite surface, is obtained as

$$\frac{D}{D_0} = \frac{1}{1 + bq} \tag{2.123}$$

where D_0 is the dispersion as the loading $q \to 0$. Dorling et al. (1971) prepared Pt/SiO_2 catalysts by impregnation of silica with chloroplatinic acid. In agreement with discussion in section 2.4.1, since chloroplatinic acid does not interact with silica, a nonlinear dependence of surface area on loading was found as shown graphically in Figure 2.18. On the other hand, when an amine platinum complex which interacts with silica was used, a linear dependence was found for catalysts containing up to 4.5 wt% Pt.

Basset et al. (1975) and Szegner et al. (1997) used chloroplatinic acid for γ-Al_2O_3 impregnation, and found a linear dependence of catalyst surface area

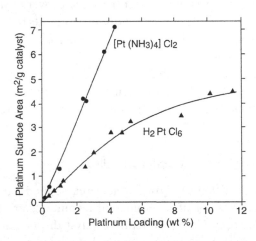

Figure 2.18. Variation of platinum surface area with platinum content for Pt/SiO_2 catalysts. The catalysts were prepared by impregnation with H_2PtCl_6 and $[Pt(NH_3)_4]Cl_2$. Solid line is equation (2.122) with $p = 84\,m^2/g$, $b = 10$. (From Dorling et al., 1971.)

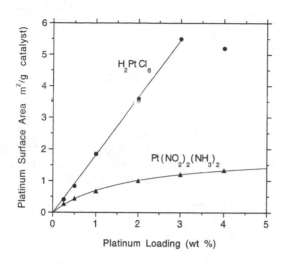

Figure 2.19. Variation of platinum surface area with platinum content for Pd/γ-Al$_2$O$_3$ catalysts. The catalysts were prepared by impregnation with H$_2$PtCl$_6$ and Pt(NO$_2$)$_2$(NH$_3$)$_2$. Solid line is equation (2.122) with $p = 109$ m^2/g, $b = 58$. (From Basset et al., 1975.)

on loading, because in this case strong precursor–support interactions exist. Both researchers observed a linear dependence up to 3 wt% Pt as shown in Figure 2.19. Beyond this, the platinum surface area remained constant, at ∼5 m^2/g, presumably due to full occupation of the adsorption sites. Further, Szegner et al. (1997) found that addition of tin to these catalysts resulted in no significant change in the accessible platinum atoms or dispersion. Basset et al. (1975) also prepared catalysts using Pt(NO)$_2$(NH$_3$)$_2$, and these showed a nonlinear relationship between surface area and loading as demonstrated in Figure 2.19. For the catalysts described above, the plot of active surface area vs catalyst loading follows equation (2.122), as shown by the corresponding solid curves in Figures 2.18 and 2.19.

Similar behavior is also exhibited by other catalytic systems. For example, Moss et al. (1979) prepared palladium catalysts by impregnating silica with tetraamminepalladous nitrate. The metal surface area measured by CO chemisorption is shown in Figure 2.20. Results for Ag/α-Al$_2$O$_3$ catalysts prepared by silver lactate

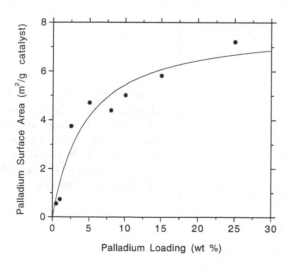

Figure 2.20. Variation of palladium surface area with palladium content for Pd/SiO$_2$ catalysts. The catalysts were prepared by impregnation with Pd(NH$_3$)$_4$(NO$_3$)$_2$. Solid line is equation (2.122) with $p = 177$ m^2/g, $b = 22$. (From Moss et al., 1979.)

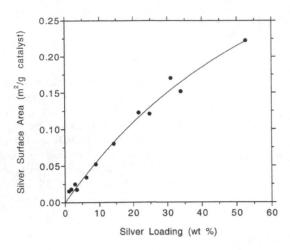

Figure 2.21. Variation of silver surface area with silver content for Ag/α-Al$_2$O$_3$ catalysts. The catalysts were prepared by impregnation with silver lactate. Solid line is equation (2.122) with $p = 0.66\,\mathrm{m^2/g}$, $b = 1.08$. (From Gavriilidis et al., 1993b.)

impregnation (Gavriilidis et al., 1993b) are shown in Figure 2.21. The nonlinear dependence of surface area on loading is due to the fact that α-Al$_2$O$_3$ is an inert support. In Figure 2.22, catalyst dispersion measured by hydrogen chemisorption as a function of loading is shown for Rh supported on SiO$_2$. The catalysts were prepared by incipient wetness impregnation of SiO$_2$, using aqueous solutions of Rh(NO$_3$)$_3$ (Underwood and Bell, 1987).

2.5   Optimal Distribution of Catalyst Loading

Since catalyst loading is the quantity controlled during catalyst preparation, it is preferable to use as the optimization variable. The focus is on reactions which are essentially structure-insensitive, i.e. for which we can assume that the rate of reaction r depends linearly on the active element surface area A:

$$r = r'A \tag{2.124}$$

where r is the reaction rate per unit weight of the catalyst pellet, and r' is the specific reaction rate per unit surface area of active element. By using equation

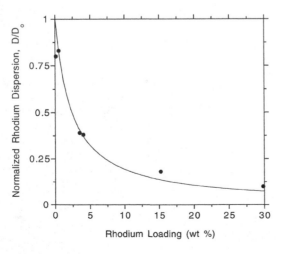

Figure 2.22. Variation of rhodium dispersion with rhodium content for Rh/SiO$_2$ catalyst. The catalysts were prepared by incipient-wetness impregnation with Rh(NO$_3$)$_3$. Solid line is equation (2.123) with $b = 42$. (From Underwood and Bell, 1987.)

(2.122) the relation between reaction rate and catalyst loading is obtained:

$$r = r' \frac{pq}{1 + bq}$$

(2.125)

2.5.1 The Problem Formulation

With a quantitative relation between reaction rate r and catalyst concentration q available, the optimization problem can now be reformulated (Baratti et al., 1993). The aim is to identify the concentration profile $q(s)$ of active element within the support which, for a fixed total amount, maximizes a given performance index of the catalyst pellet.

Definition of optimization problem
The optimization variable in dimensionless form is

$$\mu(s) = \frac{q(s)}{\bar{q}}.$$

(2.126)

Since $\bar{q}$ is the volume-average value of $q(s)$, $\mu(s)$ must satisfy the condition

$$(n+1) \int_0^1 \mu(s) s^n \, ds = 1.$$

(2.127)

In addition, on physical grounds it is reasonable to assume that $\mu(s)$ is positive and bounded, i.e.,

$$0 \le \mu(s) < \alpha$$

(2.128)

where α is the maximum allowable value of the local active element concentration on the support.

The reaction rate can be represented using equations (2.125) and (2.126) as follows:

$$r = r_0 \frac{\mu(s)}{1 + B\mu(s)}.$$

(2.129)

The factor r_0 represents the rate of reaction in a catalyst pellet with maximum dispersion, i.e. where the active element surface area per unit catalyst weight is equal to $p\bar{q}$. The dimensionless quantity $B = b\bar{q}$ is a measure of the nonlinearity of the A-vs-q relation; $B = 0$ corresponds to a linear relation, while for increasing B the nonlinearity becomes more severe. For a fixed catalyst preparation method (i.e. fixed b value), B increases with the average active element loading of the pellet, $\bar{q}$.

As in section 2.3.2, we consider a general reacting system, where any number of reactions occur, following arbitrary kinetics, in the presence of external mass and heat transport resistances. The mass and energy balances in matrix form are

$$L[\mathbf{x}] = \frac{1}{s^n} \frac{d}{ds} \left(s^n \frac{d\mathbf{x}}{ds} \right) = \frac{\mu(s)}{1 + B\mu(s)} \mathbf{w}(\mathbf{x})$$

(2.130)

with BCs

$$s = 0: \quad \frac{d\mathbf{x}}{ds} = \mathbf{0} \tag{2.131a}$$

$$s = 1: \quad \frac{d\mathbf{x}}{ds} = \mathbf{A}(\mathbf{x}_f - \mathbf{x}) \tag{2.131b}$$

where the vectors $\mathbf{x}$, $\mathbf{x}_f$, and $\mathbf{w}$ and matrix $\mathbf{A}$ are defined in (2.101) and (2.102). The effectiveness factor η_M is defined as the ratio between the actual rate of consumption of reactant M and the corresponding rate in the absence of all transport resistances in a catalyst pellet with the same overall loading of active element distributed *uniformly* throughout the support [i.e. $\mu(s) = 1$]. Note that the pellet used as reference in this definition is uniform while that under examination is nonuniform, so they may have different values of the overall active surface area. Thus

$$\eta_M = \frac{(n+1)(1+B)}{D_M \sum_{j=1}^{J} \nu_{Mj} \phi_j^2} \int_0^1 \frac{\mu(s)}{1 + B\mu(s)} w_M(\mathbf{x}) s^n \, ds. \tag{2.132}$$

Similarly, the selectivity and yield are defined by

$$S_{NM} = \frac{D_M}{D_N} \frac{\int_0^1 \dfrac{\mu(s)}{1 + B\mu(s)} w_N(\mathbf{x}) s^n \, ds}{\int_0^1 \dfrac{\mu(s)}{1 + B\mu(s)} w_M(\mathbf{x}) s^n \, ds} \tag{2.133}$$

and

$$Y_{NM} = \eta_M S_{NM} = \frac{(n+1)(1+B)}{D_N \sum_{j=1}^{J} \nu_{Mj} \phi_j^2} \int_0^1 \frac{\mu(s)}{1 + B\mu(s)} w_N(\mathbf{x}) s^n \, ds. \tag{2.134}$$

Thus, summarizing, the optimization problem consists in evaluating the function $\mu(s)$, which maximizes one of the performance indexes above, under the constraints of the diffusion–reaction equations (2.130)–(2.131) and of equations (2.127) and (2.128).

Development of the Hamiltonian

The optimization procedure is based on the variational method. As described in Appendix A, this requires derivation of the relevant Hamiltonian which is a function providing the overall effect of changes in the optimization function. For this, it is convenient to rewrite the original problem (2.130) by introducing the new variables γ and h as follows:

$$\frac{d\mathbf{x}}{ds} = \gamma s^{-n} \tag{2.135}$$

$$\frac{d\gamma}{ds} = \frac{\mu(s)}{1 + B\mu(s)} \mathbf{w}(\mathbf{x}) s^n \tag{2.136}$$

$$\frac{dh}{ds} = \mu(s) s^n \tag{2.137}$$

where the last equation arises from the integral constraint (2.127). The corresponding BCs follow from equations (2.127), (2.131), and (2.135):

$$s = 0: \qquad \gamma = 0 \tag{2.138a}$$

$$s = 1, \qquad \mathbf{\Lambda(v_I \cdot v)} \tag{2.138b}$$

$$s = 0: \qquad h = 0 \tag{2.138c}$$

$$s = 1: \qquad h = \frac{1}{n+1}. \tag{2.138d}$$

When the effectiveness factor (2.132) is the objective functional, the Hamiltonian is defined by (cf. Appendix A)

$$H = \frac{\mu(s)}{1 + B\mu(s)}\left[w_M(\mathbf{x}) + \boldsymbol{\lambda}_2^T\mathbf{w}(\mathbf{x})\right]s^n + \boldsymbol{\lambda}_1^T\boldsymbol{\gamma}s^{-n} + \lambda_3\mu(s)s^n \tag{2.139}$$

where $\boldsymbol{\lambda}_1, \boldsymbol{\lambda}_2$ (column vectors, with $I+1$ elements), and λ_3 are the Lagrange multipliers defined by the following differential equations:

$$\frac{d\boldsymbol{\lambda}_1}{ds} = -\frac{\mu(s)}{1 + B\mu(s)}s^n[\mathbf{w}_{Mx} + \mathbf{J}^T\boldsymbol{\lambda}_2] \tag{2.140}$$

$$\frac{d\boldsymbol{\lambda}_2}{ds} = -\boldsymbol{\lambda}_1 s^{-n} \tag{2.141}$$

$$\frac{d\lambda_3}{ds} = 0 \tag{2.142}$$

where $\mathbf{w}_{Mx}$ is the column vector of derivatives of w_M with respect to $\mathbf{x}$, and $\mathbf{J}$ is the Jacobian matrix of vector $\mathbf{w}$ with respect to the independent variable $\mathbf{x}$:

$$\mathbf{w}_{Mx} = \frac{\partial w_M}{\partial \mathbf{x}} = \begin{bmatrix} \partial w_M/\partial x_1 \\ \partial w_M/\partial x_2 \\ \vdots \\ \partial w_M/\partial x_1 \\ \partial w_M/\partial x_{I+1} \end{bmatrix} \tag{2.143}$$

$$\mathbf{J} = \begin{bmatrix} \partial w_1/\partial x_1 & \partial w_1/\partial x_2 & \cdots & \partial w_1/\partial x_I & \partial w_1/\partial x_{I+1} \\ \partial w_2/\partial x_1 & \partial w_2/\partial x_2 & \cdots & \partial w_2/\partial x_I & \partial w_2/\partial x_{I+1} \\ \vdots & \vdots & & \vdots & \vdots \\ \partial w_I/\partial x_1 & \partial w_I/\partial x_2 & \cdots & \partial w_I/\partial x_I & \partial w_I/\partial x_{I+1} \\ \partial w_{I+1}/\partial x_1 & \partial w_{I+1}/\partial x_2 & \cdots & \partial w_{I+1}/\partial x_I & \partial w_{I+1}/\partial x_{I+1} \end{bmatrix}. \tag{2.144}$$

The relevant BCs are

$$s = 0: \qquad \boldsymbol{\lambda}_1 = 0 \tag{2.145a}$$

$$s = 1: \qquad \boldsymbol{\lambda}_1 = \mathbf{A}^T\boldsymbol{\lambda}_2. \tag{2.145b}$$

When the yield Y_{NM} is considered as the objective function, the same expressions for the Hamiltonian and the Lagrange multipliers are obtained, but $w_M(\mathbf{x})$ is now replaced by $w_N(\mathbf{x})$ (cf. Appendix A).

In the case where the objective functional is the selectivity S_{NM}, it is convenient to introduce two auxiliary variables x_N and x_M defined as follows:

$$\frac{dx_N}{ds} = \frac{\mu(s)}{1 + B\mu(s)} w_N(\mathbf{x})s^n \tag{2.146}$$

$$\frac{dx_M}{ds} = \frac{\mu(s)}{1 + B\mu(s)} w_M(\mathbf{x})s^n \tag{2.147}$$

with initial conditions

$$s = 0: \qquad x_N = 0 \tag{2.148a}$$

$$s = 0: \qquad x_M = 0 \tag{2.148b}$$

Using the above equations, the selectivity definition (2.133) becomes

$$S_{NM} = \frac{D_M x_N(1)}{D_N x_M(1)} \tag{2.149}$$

The variational problem now consists of the objective functional (2.149), where the state variables satisfy the differential equations (2.135)–(2.137), (2.146), and (2.147), with BCs (2.138) and (2.148). The relevant Hamiltonian is (cf. Appendix A)

$$H = \frac{\mu(s)}{1 + B\mu(s)} \left[\lambda_2^T \mathbf{w}(\mathbf{x}) + \lambda_4 w_N(\mathbf{x}) + \lambda_5 w_M(\mathbf{x}) \right] s^n + \lambda_1^T \gamma s^{-n} + \lambda_3 \mu(s) s^n \tag{2.150}$$

where the Lagrange multipliers are defined by the differential equations (2.141) and (2.142) together with

$$\frac{d\lambda_1}{ds} = -\frac{\mu(s)}{1 + B\mu(s)} s^n (\mathbf{J}^T \lambda_2 + \lambda_4 \mathbf{w}_{Nx} + \lambda_5 \mathbf{w}_{Mx}) \tag{2.151}$$

$$\frac{d\lambda_4}{ds} = 0 \tag{2.152}$$

$$\frac{d\lambda_5}{ds} = 0 \tag{2.153}$$

along with BCs (2.145) and

$$s = 1: \qquad \lambda_4 = \frac{D_M}{D_N x_M} \tag{2.154}$$

$$s = 1: \qquad \lambda_5 = -\frac{D_M x_N}{D_N x_M^2}. \tag{2.155}$$

As explained in Appendix A, the solution of the optimization problem is given by the active element concentration $\mu(s)$ which maximizes the Hamiltonian H defined by equation (2.139) or (2.150), depending on the specific performance index considered as the objective functional, while simultaneously allowing for the constraint (2.128). The solution is described below for four cases (Baratti et al.,

1993, 1997). The first is the case of a single first-order isothermal reaction where an analytical solution can be obtained, the second the case of linear dependence between the active element surface area and its concentration, and the third the case of a single first-order nonisothermal reaction where a numerical method of general applicability is used for the solution. In the fourth case, the same method is applied to investigate complex reacting systems which give rise to unusual optimal loading distributions.

2.5.2 A Single First-Order Isothermal Reaction

For the case of a single first-order isothermal reaction, the system of differential equations (2.130)–(2.131) reduces to

$$L[u] = \phi^2 \frac{\mu(s)}{1 + B\mu(s)} u \tag{2.156}$$

with BCs

$$s = 0: \qquad \frac{du}{ds} = 0 \tag{2.157a}$$

$$s = 1: \qquad u = 1 \tag{2.157b}$$

where interphase mass transfer resistance has been neglected. Following the general procedure outlined above, the original problem (2.156) can be rewritten as follows:

$$\frac{du}{ds} = \gamma s^{-n} \tag{2.158}$$

$$\frac{d\gamma}{ds} = \phi^2 \frac{\mu(s)}{1 + B\mu(s)} s^n u \tag{2.159}$$

$$\frac{dh}{ds} = \mu(s)s^n \tag{2.160}$$

with the corresponding conditions

$$s = 0: \qquad \gamma = 0 \tag{2.161a}$$
$$s = 1: \qquad u = 1 \tag{2.161b}$$
$$s = 0: \qquad h = 0 \tag{2.161c}$$
$$s = 1: \qquad h = \frac{1}{n+1}. \tag{2.161d}$$

Using the effectiveness factor as the objective functional, the Hamiltonian is given by (2.139):

$$H = \frac{\mu(s)}{1 + B\mu(s)} (1 + \lambda_2)\phi^2 u s^n + \lambda_1 \gamma s^{-n} + \lambda_3 \mu(s)s^n \tag{2.162}$$

and the Lagrange multipliers by (2.140)–(2.142):

$$\frac{d\lambda_1}{ds} = -\frac{\mu(s)}{1 + B\mu(s)}(1 + \lambda_2)\phi^2 s^n \tag{2.163}$$

$$\frac{d\lambda_2}{ds} = -\lambda_1 s^{-n} \tag{2.164}$$

$$\frac{d\lambda_3}{ds} = 0 \tag{2.165}$$

with BCs

$$s = 0: \qquad \lambda_1 = 0 \tag{2.166a}$$
$$s = 1: \qquad \lambda_2 = 0. \tag{2.166b}$$

According to the maximum principle, in the case where the optimization function is not bounded [i.e., the constraint (2.128) is not included, or $\alpha = \infty$], the necessary condition for the Hamiltonian H to be maximum leads to

$$\frac{\partial H}{\partial \mu} = \frac{1}{[1 + B\mu(s)]^2}(1 + \lambda_2)\phi^2 u s^n + \lambda_3 s^n = 0. \tag{2.167}$$

By comparing equations (2.158) and (2.159) with equations (2.163) and (2.164), together with the corresponding BCs (2.161a), (2.161b) and (2.166a), (2.166b), we obtain

$$u = 1 + \lambda_2 \tag{2.168}$$

$$\gamma = -\lambda_1 \tag{2.169}$$

which may be verified by direct substitution. Using equation (2.168), the optimality condition (2.167) reduces to

$$\frac{u}{1 + B\mu(s)} = \frac{\sqrt{-\lambda_3}}{\phi} \tag{2.170}$$

which, substituted in equations (2.158)–(2.160), leads to a system of three ODEs with four BCs (2.161a–d). This set can be solved analytically to yield the three functions $u(s)$, $\gamma(s)$, and $h(s)$ in addition to λ_3, which according to equation (2.165) is constant. The analytical expressions for the optimal active element distribution $\mu(s)$, the reactant concentration profile $u(s)$, and the catalyst pellet effectiveness factor η are reported in Table 2.4 for three pellet geometries: slab ($n = 0$), cylinder ($n = 1$), and sphere ($n = 2$).

In Figure 2.23.a, the optimal distribution of the active element $\mu(s)$ is shown for a sphere and for various values of the dimensionless parameter $\psi = \phi/\sqrt{B}$, which can be regarded as a modified Thiele modulus that takes into account the nonlinearity in the dependence of the active surface area upon the active element concentration. Note that this is the only variable needed to fully characterize the optimal distribution. For values of ψ approaching 0, the optimal distribution approaches the uniform distribution. This is physically sound, considering that when the concentration of the reactant inside the pellet is uniform (small ϕ values) or

Table 2.4. Optimal active element distribution, with the corresponding reactant concentration profile and the effectiveness factor, for three pellet geometries and for a single first-order isothermal reaction ($\psi = \phi/\sqrt{B}$).

n	Active element distribution $\mu(s)$	Reactant concentration $u(s)$	Effectiveness factor $\eta = (n+1)(1+B)\tau$
0	$\psi \dfrac{\cosh \psi s}{\sinh \psi}$	$(1-\tau)\dfrac{\cosh \psi s}{\cosh \psi} + \tau$	$\tau = \dfrac{\tanh \psi}{\tanh \psi + \psi B}$
1	$\dfrac{\psi}{2}\dfrac{I_0(\psi s)}{I_1(\psi)}$	$(1-\tau)\dfrac{I_0(\psi s)}{I_0(\psi)} + \tau$	$\tau = \dfrac{I_1(\psi)}{\frac{B\psi}{2}I_0(\psi) + I_1(\psi)}$
2	$\dfrac{\psi^2}{3s}\dfrac{\sinh \psi s}{\psi \cosh \psi - \sinh \psi}$	$\dfrac{1-\tau}{s}\dfrac{\sinh \psi s}{\sinh \psi} + \tau$	$\tau = \dfrac{\psi - \tanh \psi}{(\frac{B\psi^2}{3} - 1)\tanh \psi + \psi}$

when the nonlinearity in the A-vs-q dependence is strong (large B values), the catalyst is more effectively utilized when it is distributed as uniformly as possible. On the other hand, when $\psi \to \infty$, the optimal distribution becomes a progressively steeper monotonically increasing function of s, exhibiting its maximum value at the pellet external surface. This occurs when either $B \to 0$ or $\phi \to \infty$. For $B \to 0$, since there is no problem of loss of catalyst surface area with increasing loading, it is better to locate all the catalyst where the reactant concentration is maximum, i.e. at the external surface of the pellet. For increasing values of ϕ, no matter how large B is, the reaction takes place in a narrower region near the pellets external surface, which is the only place where the active element can be profitably located.

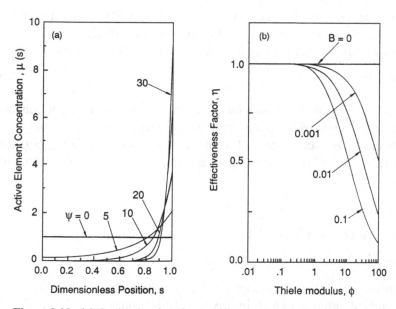

Figure 2.23. (a) Optimal active element distribution for various values of the modified Thiele modulus ψ, and (b) corresponding values of the effectiveness factor. Reaction system: a single isothermal first-order reaction in a sphere ($n = 2$). (From Baratti et al., 1993.)

This can be seen explicitly from the expressions for the optimal $\mu(s)$ reported in Table 2.4, where for any of the examined geometries, in the limit as $\psi \to \infty$,

$$\mu(s) \to \frac{\psi}{(n+1)s^{n/2}} \exp[-\psi(1-s)]. \tag{2.171}$$

Hence, in the limit of $\psi \to \infty$, $\mu(s)$ approaches a Dirac-delta function located at the pellet's external surface. This result is consistent with all the previous cases examined, where a linear A-vs-q dependence was implicitly assumed.

The corresponding optimal values of the effectiveness factor η are shown in Figure 2.23.b as a function of the Thiele modulus ϕ, for various values of B. For $B = 0$, the optimal catalyst distribution is a Dirac-delta function located at the pellet external surface. In this case, the effectiveness factor is 1 regardless of the value of the Thiele modulus. However, for increasing B values, η decreases, because the catalyst has to be spread out towards the pellet center for optimal utilization. The decrease is more significant at large Thiele modulus. Indeed, for finite B and $\phi \to \infty$, using the relationships reported in Table 2.4 it can be shown that

$$\eta \to \frac{(n+1)(1+B)}{\phi\sqrt{B}} \tag{2.172}$$

so that the η-vs-ϕ dependence is similar to that for uniform pellets (cf. Aris, 1975).

2.5.3 Linear Dependence between the Active Element Surface Area and Its Loading

When the dependence between the active element surface area and its loading is linear, i.e. $B = 0$, the expressions of both Hamiltonians (2.139) and (2.150) are linear with respect to μ, and can be reduced to the following form:

$$H = \Phi\mu + \omega \tag{2.173}$$

The function Φ is called *switching function*, and its expression depends on the reaction system under consideration as well as the performance index to be optimized. Both Φ and ω are independent of μ. In particular, considering the performance indices (2.132)–(2.134), the following expressions for the switching function are obtained:

$$\Phi = \left[w_M(\mathbf{x}) + \lambda_2^T\mathbf{w}(\mathbf{x}) + \lambda_3\right]s^n \qquad \text{for} \quad \eta_M \tag{2.174}$$

$$\Phi = \left[w_N(\mathbf{x}) + \lambda_2^T\mathbf{w}(\mathbf{x}) + \lambda_3\right]s^n \qquad \text{for} \quad Y_{NM} \tag{2.175}$$

$$\Phi = \left[\lambda_2^T\mathbf{w}(\mathbf{x}) + \lambda_3 + \lambda_4 w_N(\mathbf{x}) + \lambda_5 w_M(\mathbf{x})\right]s^n \qquad \text{for} \quad S_{NM} \tag{2.176}$$

In optimal-control theory, a system for which the Hamiltonian is a linear function of the control variable (which in this case corresponds to μ), and the latter is constrained between a minimum and a maximum setting [which corresponds to equation (2.128)], is known as a *bang–bang control system* (cf. Ray, 1981). In this case, the Hamiltonian is maximized by setting the control variable at its maximum value if the switching function is positive and at its minimum value if the switching function is negative. Hence, in general, the optimal distribution $\mu(s)$ is given by a multiple *step function*, where the positions and number of steps are controlled by the number of sign changes of the switching function Φ. Accordingly, the optimal active element distribution function profile μ is given by

$$\mu = \begin{cases} \alpha & \text{when} \quad \Phi > 0 \\ 0 & \text{when} \quad \Phi < 0. \end{cases} \tag{2.177}$$

For example, in the simple case of a single first-order isothermal reaction with the effectiveness factor as the objective functional treated in section 2.5.2, the switching function can be readily obtained from equations (2.162) and (2.168):

$$\Phi = (u^2\phi^2 + \lambda_3)s^n. \tag{2.178}$$

Since u^2 and s^n are increasing functions of s and λ_3 is constant [see equation (2.165)], Φ is a monotonically increasing function and has only one sign change for $s \in [0, 1]$. This implies that the optimal distribution is given by a single step of height α, located at the pellet surface (i.e. an eggshell distribution), whose half-width Δ satisfies the following relation derived from constraint (2.127):

$$1 - (1 - 2\Delta)^{n+1} = \frac{1}{\alpha}. \tag{2.179}$$

It can be readily seen that in the case where the optimal distribution is not bounded from above, i.e. $\alpha \to \infty$, then $\Delta \to 0$ and the distribution approaches a Dirac-delta located at $s = 1$. Thus, the Dirac-delta is in fact a special case (for $\alpha \to \infty$) of the optimal distribution, while the step distribution is optimal for finite α.

For other reacting systems, the expression for the switching function Φ, whose sign changes dictate the number of steps of the optimal distribution, will be different. Insight can be gained by studying the switching function, realizing that for complex reaction systems this study can become cumbersome. In this case, the numerical method presented in the next section may provide a more convenient alternative.

2.5.4 First-Order Nonisothermal Reactions: Numerical Optimization

In sections 2.5.2 and 2.5.3, two situations were examined where the optimal catalyst distribution was obtained analytically. However, in many cases it is more efficient to solve the optimization problem directly through a suitable numerical technique, which is presented next. Using the orthogonal collocation method (cf. Finlayson, 1980; Villadsen and Michelsen, 1978), the diffusion–reaction differential equations reduce to a set of nonlinear algebraic equations, whose solution is obtained through the Newton–Raphson method. The values of the active element concentration at the collocation points are regarded as the adjustable parameters of the optimization problem. The performance index of interest is evaluated using Gauss–Jacobi quadrature formulae, where the quadrature points coincide with the collocation points (Villadsen and Michelsen, 1978).

The optimal catalyst distribution can be a *discontinuous* function. For example, in section 2.5.3 (where $B = 0$) it is a step function. Based on continuity arguments, it is reasonable to expect that for nonzero but small B values the optimal distribution will be different from a step distribution but retain the feature of being nonzero only in a small interval of the domain $0 \leq s \leq 1$ and zero outside, while exhibiting discontinuities at the interval boundaries. In such a case, the optimal solution can be obtained conveniently by using orthogonal collocation on finite elements, where the boundaries are located where the discontinuities arise. However, such locations are not known *a priori*, and therefore the numerical determination of the optimal distribution becomes difficult. For this purpose, during optimization the elements are not fixed but have moving boundaries located at the discontinuity points.

Specifically, the optimization is carried out as follows:

1. The optimization problem is solved using orthogonal collocation applied to the entire domain $0 \leq s \leq 1$ with an increasing number of discretization points (typically 4 to 7).

2. If the obtained solution exhibits regions where $\mu = 0$, then orthogonal collocation on finite elements is applied, by placing the element boundaries so that each region where $\mu = 0$ constitutes a single element. In order to allow for discontinuities, μ is allowed to take two different values at the element boundaries, depending on which element it belongs to.

3. The above procedure is iterated, by moving the location of the element boundaries until in each element the optimal solution μ is either always zero or always nonzero.

4. Convergence of the procedure is checked by verifying that the calculated optimal distribution does not change on increasing the number of collocation points in each element.

A single first-order reaction

The optimal catalyst distributions computed for a single first-order nonisothermal reaction and $B \leq 0.1$ are shown in Figure 2.24. Note that these values of B imply volume-average catalyst loadings less than 1 wt% for the Pt/SiO$_2$ system in Figure 2.18, 0.5 wt% for the Pd/SiO$_2$ system in Figure 2.20, and 0.25 wt% for the Rh/SiO$_2$ system in Figure 2.22, and thus cover the range of interest in most practical applications. For low B values, the optimal distribution exhibits some discontinuity points and strongly resembles a step distribution. For increasing values of B, high local catalyst concentration would result in significant loss of active surface area. This provides a motivation for spreading the active element. However, the latter is counterbalanced by the effect of the reactant concentration and temperature gradients, which require all the catalyst to be located where the reaction rate is

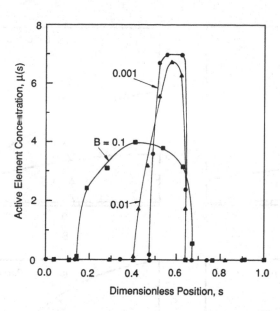

Figure 2.24. Optimal active element distribution for $\alpha = 7$ and various values of B. Reaction system: a single nonisothermal, first-order reaction, with $\beta = 0.2, \gamma = 18, \phi = 1, \mathrm{Bi_m} = \infty$, $\mathrm{Bi_h} = \infty$, and $n = 2$. (From Baratti et al., 1993.)

maximum. As a consequence, the values of the effectiveness factor corresponding to the optimal active element distribution decrease for increasing values of B, as shown in Figure 2.25. For comparison, the values of the effectiveness factor corresponding to the *uniform* distribution are also shown. These conclusions are expected to be of general validity, and in fact have also been reached for the case of bimolecular Langmuir kinetics (Baratti et al., 1993).

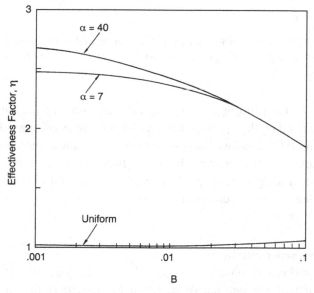

Figure 2.25. Optimal values of the effectiveness factor as a function of B. Reaction system: a single nonisothermal, first-order reaction, with $\beta = 0.2, \gamma = 18, \phi = 5, \mathrm{Bi_m} = \infty$, $\mathrm{Bi_h} = \infty$, and $n = 2$. (From Baratti et al., 1993.)

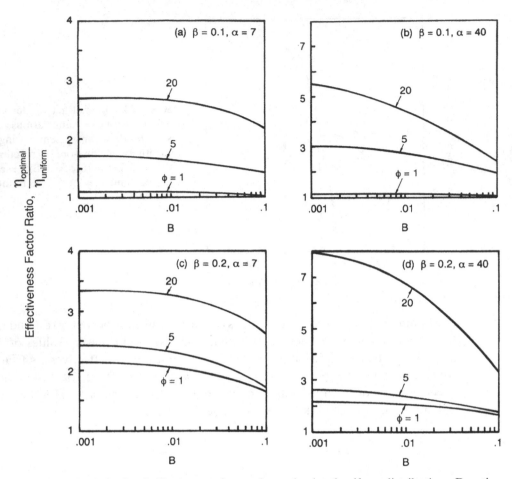

Figure 2.26. Ratio of effectiveness factors for optimal and uniform distributions. Reaction system: a single, nonisothermal, first-order reaction, with $\gamma = 18$, $Bi_m = \infty$, $Bi_h = \infty$, and $n = 2$. (From Baratti et al., 1993.)

The effectiveness factor ratio, $\eta_{optimal}/\eta_{uniform}$, is a measure of the improvement that can be obtained by the optimal catalyst distribution as compared to the uniform catalyst. This ratio is shown for various selected combinations of β, α, and ϕ values in Figure 2.26. It is evident that *the optimal distribution provides a substantial improvement* in the catalyst performance, as compared to the uniform distribution. The extent of improvement increases as β, α, or ϕ increases, and is larger for smaller B values.

Two consecutive first-order reactions

The case of two nonisothermal consecutive first-order reactions is shown in Figure 2.27. It may be seen that the optimal distribution for *selectivity* maximization remains a step distribution even for large values of B, while the optimal distribution for *effectiveness factor* maximization spreads out with increasing B. Such behavior can be justified by considering that in the case of selectivity optimization, there

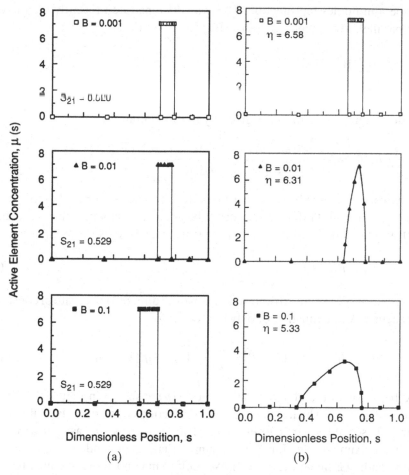

Figure 2.27. Optimal active-element distributions for $\alpha = 7$ and various values of B, in the case of (a) selectivity and (b) effectiveness factor optimization. Reaction system: two nonisothermal, first-order, consecutive reactions, with $\beta_1 = 0.2$, $\beta_2 = 0.1$, $\gamma_1 = 20$, $\gamma_2 = 10$, $\phi = 1$, $D_1 = D_2 = 1$, $Bi_{m1} = Bi_{m2} = \infty$, $Bi_h = \infty$, and $n = 2$. (From Baratti et al., 1993.)

is no issue of performance loss owing to poor catalyst utilization in terms of the active surface area; the key parameter is the ratio of the two reaction rates. This is not the case for performance indices based on catalyst productivity, which lead to broader optimal distributions with increasing values of B. Thus, we have the important conclusion that *even for the same reaction network and physicochemical parameters, the optimal catalyst distribution depends on the choice of the catalyst performance index selected for optimization.*

2.5.5   Multistep Optimal Loading Distribution

Even in the simple case where the catalyst surface area depends linearly on loading (section 2.5.3), the theoretical result indicated that the optimal catalyst distribution can be a multiple-step function. This was demonstrated recently for the case of two parallel, irreversible reactions (Baratti et al., 1997)

$$A \xrightarrow{1} B$$
$$A \xrightarrow{2} C$$

(2.180)

whose dimensionless rates are given by the Langmuir form

$$f_j(u, \theta) = \frac{u^{m_j} \exp[\gamma_j (1 - 1/\theta)]}{\{1 + \sigma_j u \exp[\varepsilon_j (1 - 1/\theta)]\}^2}, \qquad j = 1, 2.$$

(2.181)

In the case of negligible external mass and heat transfer resistances and equal heats of reaction, Prater's relationship between the dimensionless reactant concentration u and dimensionless temperature θ leads to [see also equation (2.43)]

$$\theta = 1 + \beta(1 - u).$$

(2.182)

This relation allows one to represent the overall dimensionless rate of consumption of reactant A as a function of only one variable:

$$f(u) = f_1[u, 1 + \beta(1 - u)] + \frac{k_{2,\mathrm{f}}}{k_{1,\mathrm{f}}} f_2[u, 1 + \beta(1 - u)].$$

(2.183)

The behavior of the global reaction rate $f(u)$ is shown in Figure 2.28 for three values of $k_{2,\mathrm{f}}/k_{1,\mathrm{f}}$, which is the ratio of the reaction constants at bulk fluid temperature. These values were chosen for illustrative purposes, so as to clearly separate the two maxima of function f. It is assumed that the dependence of catalyst activity on loading is linear (i.e. $B = 0$). We seek to maximize the overall effectiveness factor

$$\eta = \frac{n+1}{f(1)} \int_0^1 \mu(s) f(u) s^n \, ds$$

(2.184)

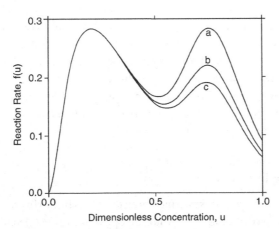

Figure 2.28. Global reaction rate $f(u)$ as a function of reactant concentration u for various $k_{2,\mathrm{f}}/k_{1,\mathrm{f}}$ values: curve a, 0.08; curve b, 0.06; curve c, 0.05. Reaction system: parallel nonisothermal reactions following Langmuir–Hinshelwood kinetics, according to equation (2.181); $\beta_1 = 0.3$, $\beta_2 = 0.3$, $\gamma_1 = 20$, $\gamma_2 = 35$, $\sigma_1 = 8$, $\sigma_2 = 0.02$, $\varepsilon_1 = 0$, $\varepsilon_2 = 50$, $m_1 = m_2 = 2$. (From Baratti et al., 1997.)

while satisfying the constraints (2.127) and (2.128). Equations (2.135)–(2.137) now become

$$\frac{du}{ds} = \gamma s^{-n} \tag{2.185}$$

$$\frac{d\gamma}{ds} = \psi'(u) f'(u) s^n \tag{2.186}$$

$$\frac{dh}{ds} = \mu(s)s^n. \tag{2.187}$$

The corresponding BCs are

$$s = 0: \qquad \gamma = 0 \tag{2.188a}$$
$$s = 1: \qquad u = 1 \tag{2.188b}$$
$$s = 0: \qquad h = 0 \tag{2.188c}$$
$$s = 1: \qquad h = \frac{1}{n+1}. \tag{2.188d}$$

Using the overall effectiveness factor as the objective functional, from equation (2.139) the Hamiltonian is given by

$$H = \Phi\mu(s) + \lambda_1\gamma(s)s^{-n} \tag{2.189}$$

where the switching function Φ is

$$\Phi = [(1+\lambda_2)\phi^2 f + \lambda_3]s^n. \tag{2.190}$$

Similarly, the Lagrange multipliers λ_1, λ_2, and λ_3 are defined as [cf. equations (2.140)–(2.142)]

$$\frac{d\lambda_1}{ds} = -\phi^2(1+\lambda_2)\,\mu(s)s^n\frac{\partial f}{\partial u} \tag{2.191}$$

$$\frac{d\lambda_2}{ds} = -\lambda_1 s^{-n} \tag{2.192}$$

$$\frac{d\lambda_3}{ds} = 0 \tag{2.193}$$

with the BCs

$$s = 0: \qquad \lambda_1 = 0 \tag{2.194a}$$
$$s = 1: \qquad \lambda_2 = 0. \tag{2.194b}$$

Since the Hamiltonian is linear with respect to $\mu(s)$, the optimal catalyst loading profile is the following step function (see also section 2.5.3):

$$\mu = \begin{cases} \alpha & \text{when} \quad \Phi > 0 \\ 0 & \text{when} \quad \Phi < 0. \end{cases} \tag{2.195}$$

The number and location of steps depend on the specific form of the switching function, whose evaluation requires complete solution of the problem. This has

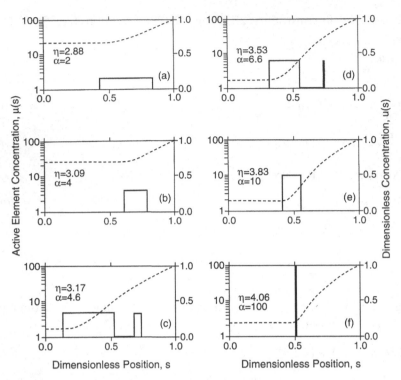

Figure 2.29. Optimal active element distribution μ (continuous curve) and reactant concentration u (dashed curve) within the catalyst pellet for various values of the upper bound α on the dimensionless active element concentration. The values of the corresponding effectiveness factors are also indicated. Reaction system: parallel nonisothermal reactions following Langmuir–Hinshelwood kinetics, according to equation (2.181); $\beta_1 = 0.3$, $\beta_2 = 0.3$, $\gamma_1 = 20$, $\gamma_2 = 35$, $\sigma_1 = 8$, $\sigma_2 = 0.02$, $\varepsilon_1 = 0$, $\varepsilon_2 = 50$, $m_1 = m_2 = 2$, $\phi = 3$, $k_{2,f}/k_{1,f} = 0.06$. (From Baratti et al., 1997.)

been carried out numerically, using the method described in section 2.5.4, and the results are shown in Figure 2.29 for a few selected values of the upper bound α on the dimensionless catalyst loading.

For low values of α, the optimal distribution is given by a single-step function. As α increases, it becomes possible to concentrate more catalyst in a given support area. In particular, for the cases shown in Figure 2.29.c and d, the catalyst has been placed in two locations, which (as can be seen from the corresponding average values of u) are near to those where $f(u)$ exhibits its maxima. In these cases, the optimal distribution is given by a *two-step function*. When α is further increased ($\alpha > 7.4$), the optimal distribution again takes the form of a single-step function, since α is now large enough so that all the catalyst can be located in the vicinity of the point where its utilization is maximized, i.e. where $f(u)$ exhibits the larger of its maxima. In the present case (see Figure 2.28), this corresponds to the lower value of u, and hence to the inner of the two steps noted above. In the limit as $\alpha \rightarrow \infty$, the optimal distribution approaches a Dirac-delta function.

In the example discussed above, the appearance of a two-step optimal catalyst distribution is due to the fact that the overall reaction rate exhibits two local maxima with respect to reactant concentrations. This observation leads to the general conclusion that, in practical applications, the optimal active catalyst distribution can exhibit the multiple-step form only in cases where the involved reactions have different kinetics. Since we are considering here the case where the catalyst is the same, this implies that the two reactions should exhibit different mechanisms, i.e. involving the breaking of different chemical bonds. Examples include isomerization and hydrogenation of an unsaturated hydrocarbon, and parallel exothermic reactions with different kinetics and activation energies (Baratti et al., 1997).

The practical implication of these results is that, particularly when dealing with complex systems, as in most industrial applications, one should not assume *a priori* that the optimal catalyst distribution has the form of a single step. It can in fact well be a multiple-step distribution.

2.6   Experimental Studies

In addition to the considerable amount of theoretical research, several experimental studies have been conducted which demonstrate the improved performance of nonuniformly distributed catalysts. In this context, a substantial volume of work is related to automotive exhaust catalysts, and will be discussed in section 6.1. The experimental research, in general, shows good qualitative, and in certain cases also quantitative, agreement with the theory.

2.6.1   Oxidation Reactions

Kotter and Riekert (1983a,b) investigated the triangular reaction network of partial oxidation of propene to acrolein on $CuO/\alpha\text{-}Al_2O_3$. Three types of catalyst were prepared, having approximately the same amount of CuO, deposited in uniform, eggshell, and egg-yolk distributions. They found that the selectivity to acrolein was higher for the eggshell catalyst and progressively lower in the uniform and egg-yolk catalysts, in good agreement with theoretical predictions for the reaction network.

On $Pt/\gamma\text{-}Al_2O_3$ catalysts prepared by coimpregnation techniques, Chemburkar (1987) studied carbon monoxide oxidation, which in excess oxygen exhibits a maximum in the rate as a function of CO concentration. Steady-state multiplicity behavior was observed, and it was found that the catalyst performance was sensitive to even small catalytic activity present in the portion of the support surrounding the active layer (see Figure 2.30). In agreement with this study, Harold and Luss (1987) demonstrated theoretically that multiplicity features for carbon monoxide oxidation are altered significantly by small variations in the activity profile of an eggshell-type catalyst.

Gavriilidis and Varma (1992) studied the ethylene epoxidation reaction network on $Ag/\alpha\text{-}Al_2O_3$ step catalyst in a single-pellet reactor with external recycle. As shown in Figure 2.31, the selectivity to ethylene oxide is higher when the active material is located as a thin layer at the external surface of the pellet. These results,

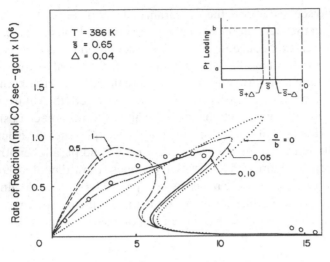

Figure 2.30. Simulations using the activity profile illustrated in the inset, for CO oxidation on Pt/γ-Al$_2$O$_3$, for various values of a/b. The circles denote experimental data. The curves represent solutions of the diffusion–reaction model. (From Chemburkar, 1987.)

which were obtained under excess oxygen are, consistent with the detrimental effect of temperature gradients on selectivity (Morbidelli et al., 1984a; Pavlou and Vayenas, 1990a), owing to the higher activation energy of ethylene combustion than of ethylene epoxidation. Under ethylene-rich conditions, in addition to temperature gradients, oxygen concentration gradients affect the selectivity negatively (Yeung et al., 1998). Reactant concentration gradients are also detrimental for

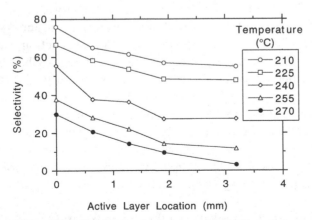

Figure 2.31. Selectivity to ethylene oxide during ethylene epoxidation on Ag/α-Al$_2$O$_3$, as a function of active layer location, for various bulk fluid temperatures; inlet ethylene concentration 7.2%. (From Gavriilidis and Varma, 1992.)

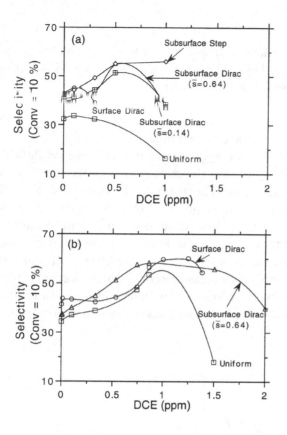

Figure 2.32. Ethylene oxide selectivity during ethylene epoxidation on Ag/α-Al$_2$O$_3$, as a function of DCE concentration at a fixed 10% ethylene conversion for various distributions of catalyst: (a) no ethane, and (b) 0.5% ethane. (From Yeung et al., 1998.)

conversion. As a result, a surface Dirac-type catalyst pellet exhibits high ethylene conversion as well as high selectivity and yield towards ethylene oxide.

When dichloroethane (DCE) is added to the feed as a promoter, the situation changes (Yeung et al., 1998). In general, subsurface catalyst distributions have the best selectivity and yield, as shown in Figure 2.32.a for 10% ethylene conversion. DCE inhibits both the epoxidation and combustion reactions. In particular, low concentrations of DCE improve selectivity, while higher concentrations are detrimental due to formation of bulk silver chloride. A surface Dirac-type catalyst has better performance at low DCE levels ($\leq$0.2 ppm), while enhanced selectivity is attained at higher DCE levels (0.7–1.5 ppm) for subsurface Dirac catalysts. The observed effects are due to transport resistances, which decrease the concentration of DCE along the depth of the pellet. An active layer placed at subsurface locations is exposed to a lower concentration of DCE than is a surface catalyst. This gives rise to the desired inhibition effect that enhances selectivity.

Experiments were also performed with ethane present in the feed. Ethane reacts with adsorbed chlorine, thus preventing its accumulation at the catalyst surface. Similar behavior was obtained (see Figure 2.32.b), but the maximum displayed by the subsurface Dirac catalyst was broader, allowing a larger window of operation and greater tolerance for fluctuations in the feed DCE concentration. Placing the active layer at a subsurface location was shown also to significantly reduce the amount of trace contaminants present in the feed gas that reach the silver catalyst.

2.6.2   Hydrogenation Reactions

Masi et al. (1988) studied ethylene hydrogenation in an external recycle reactor, on Pt/γ-Al$_2$O$_3$ catalysts prepared by coimpregnation techniques. Steady-state multiplicity behavior was observed. For a range of ethylene concentration values, larger reaction rates were obtained by subsurface than by surface distribution. This feature was attributed to the fact that the hydrogenation reaction rate exhibits a maximum as a function of ethylene concentration.

Wu et al. (1988) investigated the hydrogenation of ethylene on Pd/γ-Al$_2$O$_3$, which follows first-order kinetics, and the methanation of CO on Ni/γ-Al$_2$O$_3$, which follows bimolecular Langmuir–Hinshelwood kinetics, in a single-pellet internal recycle reactor. The pellets were prepared by pressing the active catalyst between two alumina layers, so that step distributions at any desired location could be produced. It was found for both reactions that there exists an optimal location for the active layer within the pellet (see Figure 2.33), which moves towards the external surface as the Thiele modulus increases, according to theoretical predictions.

Au et al. (1995) studied benzene hydrogenation on Ni/kieselguhr catalyst in a single-pellet diffusion reactor. The kinetics of this reaction were correlated by an

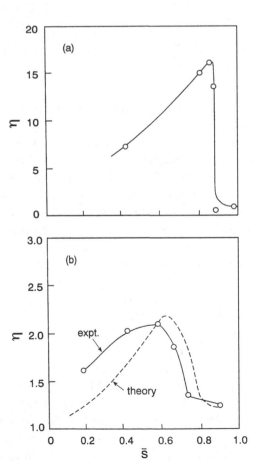

Figure 2.33. (a) Plot of experimental effectiveness factor η vs active-layer location $\bar{s}$, for ethylene hydrogenation on Pd/Al$_2$O$_3$: temperature 298 K, hydrogen concentration 5.89%, $\Delta = 0.05$. (b) Experimental and calculated (dashed line) dependences of effectiveness factor η on the active layer location $\bar{s}$ for CO methanation on Ni/Al$_2$O$_3$: temperature 533 K, CO concentration 5.0×10^{-6} mol/ml, H$_2$/CO = 7.85. (From Wu et al., 1988.)

Eley–Rideal rate equation, which is of positive order with respect to both reactants. Five different catalyst distributions ranging from preferential shell loading to preferential core loading were examined under isothermal conditions. The effectiveness factor was found to be higher for pellets with the catalyst concentrated towards the external surface, which is consistent with the Eley–Rideal kinetics. At small Thiele modulus (i.e. kinetic control), all pellets showed the same performance as expected, but the shell-type pellets showed notable improvements for larger Thiele modulus values.

In complex hydrogenation reactions, selectivity is the performance index of interest. In these cases, partially hydrogenated products are typically desired. Komiyama et al. (1997) studied liquid-phase batch hydrogenation of 1,3-butadiene over Pt/γ-Al$_2$O$_3$ catalysts, which is a triangular reaction network

$$
\begin{array}{c}
\text{Butadiene} \rightarrow \text{Butene} \\
\searrow \qquad \swarrow \\
\text{Butane}
\end{array}
\tag{2.196}
$$

where butene hydrogenation is inhibited in the presence of butadiene. Uniform, eggshell, and egg-white catalyst pellets were used, and the highest selectivities to butenes were obtained with the eggshell distribution. The residence time of butenes inside the catalyst pellet was larger for uniform and egg-white pellets, thus providing a greater chance to react further to butane. For the uniform catalyst, an interesting feature observed was that the butane fraction decreased with time during the initial stages of the reaction. This was attributed to the fact that the contribution of butene hydrogenation to butane production is larger in the beginning, due to the absence of butadiene in the pellet interior.

A similar situation exists in the case of acetylene hydrogenation on Pd/Al$_2$O$_3$. Even if ethylene is present at high concentration, 100% selectivity to ethylene can be obtained by the optimal catalyst profile (eggshell), because traces of acetylene can poison ethylene hydrogenation completely (Mars and Gorgels, 1964). Therefore, a zone in the catalyst layer which is free of acetylene is detrimental, because ethylene can be hydrogenated in this zone. For this reason, the best catalyst distribution is a sharp eggshell that minimizes acetylene concentration gradients.

Uemura and Hatate (1989) studied the same reaction on Ni/Al$_2$O$_3$ pellets with thin eggshell, thick eggshell, uniform, and egg-yolk profiles. They obtained the highest selectivity using the thin eggshell catalyst, as shown in Figure 2.34, in agreement with theoretical calculations. However, even in the absence of ethylene in the feed, the highest selectivity was about 80%, obtained at low conversion using the thin eggshell catalyst. This was attributed to the existence of a direct hydrogenation route from acetylene to ethane.

Berenblyum et al. (1986) investigated the selective hydrogenation of C$_5$–C$_8$ dienes to alkenes on eggshell Pd catalysts with different catalyst loadings (0.05–0.5 wt%) and shell thicknesses (0.1–1.5 mm). They observed that the highest diene conversion was obtained at intermediate values of catalyst loading and thickness (0.25% Pd and 0.2-mm catalyst thickness), while pellets with active layer thickness less than 0.2 mm and loadings 0.15–0.25% Pd offered the best selectivities.

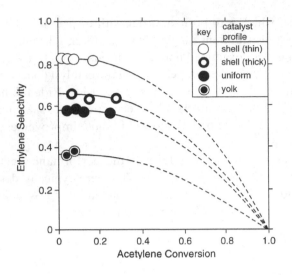

Figure 2.34. Ethylene selectivity as a function of acetylene conversion during acetylene hydrogenation on various nonuniform Ni/Al_2O_3 pellets. $T = 413$ K, $P_A = 5.1$ kPa, $P_H = 4.1$ kPa. (From Uemura and Hatate, 1989.)

This behavior was attributed to an interaction of diffusion–reaction phenomena, loading dispersion dependence, and structure sensitivity of the reactions.

2.6.3   Fischer–Tropsch Synthesis

Eggshell Co/SiO_2 catalysts were shown to be more active and selective than uniform pellets for C_5^+ production during Fischer–Tropsch synthesis (Iglesia et al., 1993; Iglesia et al., 1995; Iglesia, 1997), because they decrease diffusional restrictions that normally lead to low reaction rates and selectivities. However, a certain level of diffusional resistance was found to be beneficial to selectivity. Pellets with intermediate thickness of the catalyst layer provided maximum selectivity, because they avoid large intraphase CO concentration gradients while restricting the diffusive removal of reactive olefin products, which can continue to grow to higher molecular weight hydrocarbons. The higher C_5^+ selectivity of eggshell catalysts than of uniform ones was also reported previously in the patent literature (van Erp et al., 1986; Post and Sie, 1986).

3

Optimization of the Catalyst Distribution in a Reactor

I n single-pellet studies, it is assumed that no concentration or temperature gradients are present in the fluid phase surrounding the pellet. A reactor with external or internal recycle is one of the experimental realizations of the single-pellet concept. However, in a fixed-bed reactor, the fluid-phase composition and temperature vary with position. For this reason, the optimization problem becomes more complex. Thus, it is not surprising that relatively few reactor studies have appeared in the literature as compared with those for single pellets. In this chapter, we first discuss theoretical studies of single and multiple reactions under isothermal and nonisothermal conditions, and then present experimental work which supports the theoretical developments.

3.1 A Single Reaction

3.1.1 Isothermal Conditions

One of the earliest works in this area considered CO oxidation in excess oxygen over platinum catalyst, in monolith reactors for automobile converters (Becker and Wei, 1977a). Recall that this reaction exhibits a maximum in the rate as a function of CO concentration. The CO conversion to CO_2 as a function of the Thiele modulus (or equivalently, the catalyst temperature) for a fixed inlet CO concentration is shown in Figure 3.1 for four monoliths with different catalyst distributions (cf. Figure 2.1). The conversion of CO below 650°F was significantly higher in the inner catalyst, while above 650°F the middle catalyst performed better and attained 100% conversion. The uniform and outer catalysts also showed 100% conversion at progressively higher temperatures.

The optimization of the isothermal fixed-bed reactor, where a bimolecular Langmuir–Hinshelwood reaction occurs, was performed analytically by Morbidelli et al. (1986a,b). In this case, the mass balance for a single pellet [equation (2.2)] must be coupled with the mass balance for the fluid phase, which for a

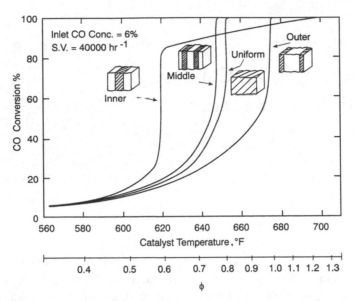

Figure 3.1. Carbon monoxide conversion to CO_2 over a mono-lith reactor with various nonuniform distributions of platinum catalyst. (From Becker and Wei, 1977a.)

heterogeneous plug-flow reactor is

$$v\frac{dC_f}{dz} = -(1-\varepsilon)\bar{r}(C_f) \tag{3.1}$$

with initial conditions

$$z = 0: \qquad C_f = C_f^0 \tag{3.2}$$

where $\bar{r}(C_f)$ represents the volume-averaged reaction rate in the catalyst phase and is expressed by

$$\bar{r}(C_f) = \frac{1}{V_p}\int_{V_p} r(C)\,a(x)\,dV_p. \tag{3.3}$$

The dimensionless form of equation (3.1) is

$$\frac{dg}{dy} = -\mathrm{Da}\,(n+1)\int_0^1 f(u)\,a(s, y)s^n\,ds \tag{3.4}$$

with initial conditions

$$y = 0: \qquad g = 1 \tag{3.5}$$

where the following dimensionless quantities have been introduced:

$$g = C_f/C_f^0, \qquad y = z/L, \qquad \mathrm{Da} = L(1-\varepsilon)r(C_f^0)/vC_f^0$$

$$u = C/C_f^0, \qquad \sigma = KC_f^0, \qquad f(u) = \frac{r(C)}{r(C_f^0)} = \frac{(1+\sigma)^2 u}{(1+\sigma u)^2}. \tag{3.6}$$

The fluid-phase balance has to be solved in conjunction with the solid-phase balance [equations (2.2) and (2.5)], which using the above dimensionless variables reduces to

$$\frac{1}{s^n}\frac{d}{ds}\left(s^n\frac{du}{ds}\right) = \phi^2 a(s) f(u) \tag{3.7}$$

with BCs

$$s = 0: \qquad \frac{du}{ds} = 0 \tag{3.8a}$$

$$s = 1: \qquad u = g \tag{3.8b}$$

where the second BC allows for the fact that the fluid-phase composition changes along the reactor length.

The objective of reactor optimization is to maximize the conversion at the reactor outlet,

$$X = 1 - g(1). \tag{3.9}$$

The conversion can be evaluated by integrating the external fluid-phase mass balance to give

$$X = \text{Da} \int_0^1 e(y)\, dy \tag{3.10}$$

where

$$e(y) = (n+1)\int_0^1 f(u)a(s, y)s^n\, ds \tag{3.11}$$

represents the contribution to reactant conversion at location y along the reactor axis. The optimization variable is the activity distribution $a(s, y)$, which now depends also on the position along the reactor axis y, because each particle along the bed may have a different distribution. Since $e(y)$ is always positive, maximum conversion is obtained by locally maximizing $e(y)$ at each $y \in [0, 1]$. Thus, the problem is reduced to optimizing the single-pellet performance at each position. The difference between the reactor and the single-pellet case is in the BC for u, which now varies with position along the reactor, i.e., $u(1, y) = g(y) \leq 1$. Using the same arguments as in the pellet case, it can be shown that the optimal distribution is a Dirac-delta function centered at a location $\bar{s}(y)$ within the pellet such that $u(\bar{s}) = u_m$, that is, where the concentration is equal to the value which maximizes the reaction rate. The method of finding the optimal catalyst location along the reactor by single-pellet optimization may be referred to as the *local* optimal catalyst distribution (OCD) approach.

Solving the pellet mass balance with the Dirac catalyst distribution

$$a(s, y) = \frac{\delta[s - \bar{s}(y)]}{(n+1)\bar{s}^n(y)} \tag{3.12}$$

Table 3.1. Location of optimal catalyst distribution along a reactor and corresponding local consumption rate for isothermal, bimolecular Langmuir–Hinshelwood kinetics, without external transport resistances.

	$\sigma < 1$ or $\sigma > 1$ and $g < 1/\sigma$	$\sigma > 1$ and $g > 1/\sigma$	
$\bar{s}_{\mathrm{opt}}(y)$	1	$1 - \Pi^{a}$	$n = 0$
		$\exp(-\Pi)$	$n = 1$
		$\dfrac{1}{1+\Pi}$	$n = 2$
$e_{\mathrm{opt}}(y)$	$f(g)$	$f(u_{\mathrm{m}}) = \dfrac{(1+\sigma)^2}{4\sigma}$	

^{a} For $\Pi > 1$, $\bar{s}_{\mathrm{opt}} = 0$ and $e_{\mathrm{opt}}(y) = f(\bar{u})$, where $\bar{u}$ is the solution of $1 - \bar{u} - \phi^2 f(\bar{u}) = 0$.

the following expressions for the optimal location are derived:

$$\bar{s}_{\mathrm{opt}}(y) = 1 - \Pi \qquad \text{for} \quad n = 0 \qquad\qquad (3.13a)$$

$$\bar{s}_{\mathrm{opt}}(y) = \exp(-\Pi) \qquad \text{for} \quad n = 1 \qquad\qquad (3.13b)$$

$$\bar{s}_{\mathrm{opt}}(y) = \frac{1}{1 + \Pi} \qquad \text{for} \quad n = 2 \qquad\qquad (3.13c)$$

where

$$\Pi = \frac{4(n+1)(\sigma g - 1)}{\phi_0^2} \qquad\qquad (3.14)$$

so that, since $g = g(y)$, $\bar{s}_{\mathrm{opt}}$ is a function of reactor position y. The resulting value of the local consumption rate, $e(y)$, is obtained by substituting the optimal Dirac distribution in equation (3.11). Note that these solutions do not hold for the cases $\sigma < 1$, or $\sigma > 1$ and $g < u_{\mathrm{m}} = 1/\sigma$. In these cases the reaction rate $f(u)$ is monotonically increasing in the interval $u \in [0, g]$ and attains its maximum value at $u = g$. Therefore, the optimal catalyst location is at the pellet external surface, i.e. $\bar{s}_{\mathrm{opt}} = 1$. Note also that for the infinite slab ($n = 0$), the value of $\bar{s}_{\mathrm{opt}}$ can become negative. In this case, as discussed in section 2.1.1, the optimum catalyst location is $\bar{s}_{\mathrm{opt}} = 0$.

The expressions for the optimal location along the reactor and the corresponding outlet reactant conversion are summarized in Table 3.1. They indicate that, starting from the reactor inlet, until the fluid reactant concentration reaches $1/\sigma$, the optimal pellets have subsurface Dirac-delta distribution. The catalyst location moves progressively towards the pellet external surface, while each pellet exhibits constant local reaction rate, e_{opt}.

When the fluid reactant concentration becomes lower than $1/\sigma$, then beyond this reactor position, the pellets have surface Dirac distribution and the local reaction rate changes with fluid reactant concentration. The fluid concentration g and the optimal location $\bar{s}_{\mathrm{opt}}$ are illustrated in Figure 3.2 for an optimal reactor

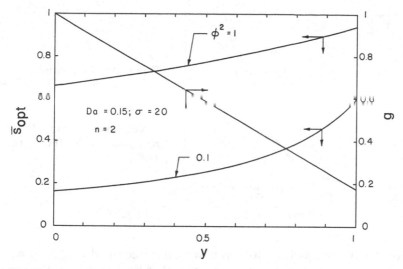

Figure 3.2. Rigorous optimal solution: profiles of the optimal active catalyst location $\bar{s}_{opt}$ and the bulk fluid concentration g along the reactor axis y for a bimolecular Langmuir–Hinshelwood reaction. (From Morbidelli et al., 1986b.)

where $g(y) > 1/\sigma$. Integrating equation (3.4) yields the fluid concentration profile

$$g(y) = 1 - \mathrm{Da}\, e_{opt} y. \tag{3.15}$$

Note that in this case, only the optimal catalyst location $\bar{s}_{opt}$ depends on the Thiele modulus, while the g profile remains unchanged, since diffusional resistances affect only the location inside the pellet where u_m is realized.

The obtained optimal solution requires the catalyst location to be different in each pellet, depending on its location along the reactor. This is difficult to realize in practice, but provides the theoretical limit of reactor performance, and hence may be referred to as the *absolute optimum* solution. The more practical situation where the catalyst location is the same for all pellets has also been examined (Lee et al., 1987a). Its performance was only slightly below the absolute optimum solution for low conversion, but became progressively worse for higher conversions. As an alternative configuration, the fixed-bed is divided into multiple zones, where in each zone the catalyst location is the same but different from the others. The optimization variables are now the catalyst location within pellets in each zone, as well as the length of the corresponding zone. As shown in Figure 3.3, for any given Damköhler number Da, and for increasing number of zones, d, the optimized reactor configuration evolves towards the absolute optimum, achieved for an infinite number of zones, $d = \infty$. By the use of only two or three zones, the theoretical optimal reactor performance is reasonably closely approached. The optimal catalyst locations and zone lengths for reactors having one, two, three, or an infinite number of zones are illustrated in Figure 3.4.

In reactor optimization studies, care should be exercised for the possible occurrence of multiple steady states. For the reaction system considered above,

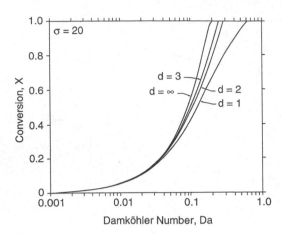

Figure 3.3. Comparison of reactor outlet conversion X for various number of zones in the reactor, d, as a function of Damköhler number Da. Bimolecular Langmuir–Hinshelwood reaction, $\sigma = 20$. (From Lee et al., 1987a.)

reactor steady-state multiplicity was observed (three steady states, of which two were stable) and attributed to the multiplicity behavior of the first layer of pellets packed along the axial direction (Morbidelli et al., 1986a). Criteria were developed to predict the existence of multiple steady states and ignition conditions. In the case of a single-zone reactor with axial dispersion, the number of possible steady states increases to nine, among which only three are stable (Lee et al., 1987b). Two sources of multiplicity are present in this case: the heterogeneous nature of the reactor and the presence of axial dispersion in the fluid phase.

Homogeneous–heterogeneous reactions

In the cases discussed up to now, only catalytic reactions have been considered. However, there are certain situations where both heterogeneous and homogeneous reactions take place, with some species participating in both types of reactions. One such case is catalytic combustion, where heterogeneous reactions can enhance homogeneous ones because the catalyst promotes the production of free radicals, which can increase the rate of homogeneous reactions (Pfefferle and Pfefferle, 1987). The simple case of two first-order reactions, one heterogeneous and the other homogeneous, was studied by Melis et al. (1997). The analysis was carried out in the absence or presence of interactions between the two reaction

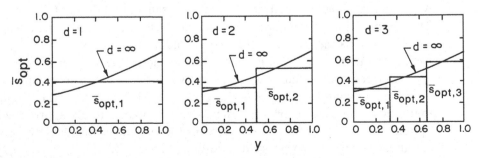

Figure 3.4. Optimal catalyst location $\bar{s}_{opt}$ along the reactor axis y for various numbers of zones in the reactor, d. Bimolecular Langmuir–Hinshelwood reaction, $\sigma = 5$, Da $= 0.3$, $\phi^2 = 0.7$, $n = 1$. (From Lee et al., 1987a.)

types, under isothermal conditions, using reactor conversion as the optimization parameter. Enhancement of homogeneous reactions was assumed to take place only in the noncatalytic portion of the reactor and was allowed for in a simple way, by a discrete increase of the apparent rate constant of the homogeneous reaction through an enhancement parameter. A zone reactor was considered, with each zone either containing catalyst or not. In catalyst-containing zones, the catalyst distribution was uniform both radially and axially.

In the absence of enhancement of homogeneous reactions, the optimal reactor was found to be a one-zone reactor containing catalyst, except for the case of negligible mass transfer resistance, where conversion was independent of catalyst distribution. When the presence of the catalyst enhanced the homogeneous reactions, it was found that a two-zone reactor was optimal, with the first zone catalytic and the second noncatalytic. When the mass transfer resistance was large and for sufficiently high values of the enhancement factor, the optimal policy was to support the homogeneous reaction to the greatest extent. This occurred with a Dirac-type first (catalytic) zone, as shown in Figure 3.5.a,b. As the mass transfer resistance decreased, as the intrinsic catalytic activity increased, or as the enhancement of homogeneous reaction became weaker, the length of the first catalytic zone increased at the expense of the noncatalytic one, which eventually disappeared, as shown in Figure 3.5.c,d, because the contribution of the heterogeneous reactions became more prominent. Even though the above study is limited to a simple case, it demonstrates that when homogeneous reactions are present the optimal catalyst distribution can be greatly affected.

3.1.2 Nonisothermal Conditions

For a single reaction taking place in an adiabatic reactor, the steady-state fluid-phase mass and energy balances for a heterogeneous plug-flow reactor are (Lee and Varma, 1987)

$$v \frac{dC_f}{dz} = -(1 - \varepsilon) \bar{r}(C_f, T_f) \tag{3.16}$$

$$\rho c_p v \frac{dT_f}{dz} = (1 - \varepsilon)(-\Delta H) \bar{r}(C_f, T_f) \tag{3.17}$$

with initial conditions

$$z = 0: \quad C_f = C_f^0, \quad T_f = T_f^0. \tag{3.18}$$

The volume-averaged reactivity of the solid catalyst phase, $\bar{r}(C_f, T_f)$, is given by equation (3.3), where now the reaction rate r is also a function of temperature. The dimensionless mass and energy balances are

$$\frac{dg}{dy} = -\text{Da}\,(n + 1) \int_0^1 f(u, \theta)\, a(s, y) s^n\, ds \tag{3.19}$$

$$\frac{d\tau}{dy} = \text{Da}\,(n + 1) \bar{\beta} \int_0^1 f(u, \theta)\, a(s, y) s^n\, ds \tag{3.20}$$

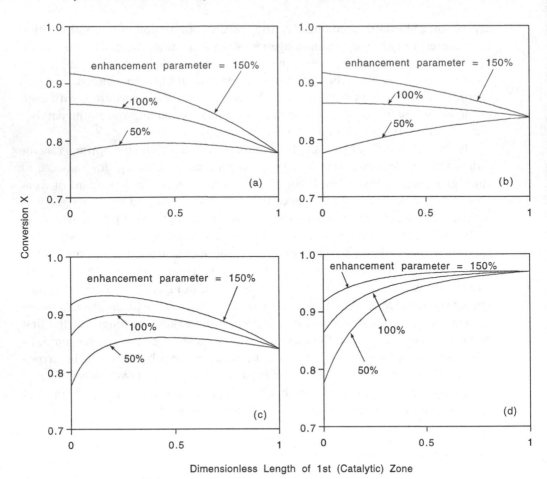

Figure 3.5. Reactor exit conversion X as a function of dimensionless length of the first (catalytic) zone when the homogeneous reaction rate in the second (noncatalytic) zone is enhanced through the catalytic reactions by the enhancement parameter shown. Da $= 1$. Dimensionless mass transfer coefficient, dimensionless heterogeneous reaction rate constant: (a) 1, 1; (b) 1, 5; (c) 5, 1; (d) 5, 5. (From Melis et al., 1997.)

with initial conditions

$$y = 0: \qquad g = 1, \quad \tau = 1 \tag{3.21}$$

where the following dimensionless quantities have been used:

$$\tau = \frac{T_f}{T_f^0}, \qquad \bar{\beta} = \frac{(-\Delta H)C_f^0}{\rho c_p\, T_f^0}, \qquad \text{Da} = \frac{L(1-\varepsilon)r\left(C_f^0, T_f^0\right)}{v C_f^0}$$

$$\theta = \frac{T}{T_f^0}, \qquad f(u,\theta) = \frac{r(C,T)}{r\left(C_f^0, T_f^0\right)}. \tag{3.22}$$

The above fluid-phase balances need to be solved in conjunction with the solid-phase balances (2.43)–(2.45), which using the above dimensionless variables take the form

$$\frac{1}{s^n}\frac{d}{ds}\left(s^n\frac{du}{ds}\right) = \phi^2 a(s, y) f(u, \theta) \tag{3.23}$$

$$\frac{1}{s^n}\frac{d}{ds}\left(s^n\frac{d\theta}{ds}\right) = -\beta\phi^2 a(s, y) f(u, \theta) \tag{3.24}$$

$$s = 0: \qquad \frac{du}{ds} = 0, \quad \frac{d\theta}{ds} = 0 \tag{3.25a}$$

$$s = 1: \qquad u = g, \quad \theta = \tau. \tag{3.25b}$$

When considering conversion optimization, it is evident from equation (3.19) that optimality is achieved by maximizing the reaction rate at all locations along the reactor. This means that, as for the isothermal reactor, optimum performance is achieved with Dirac-delta catalyst distributions centered at different locations depending on pellet position along the reactor.

The steady-state multiplicity behavior of such a reactor can be quite complex. Using singularity theory, Lee and Varma (1987) studied the multiplicity characteristics of an adiabatic plug-flow reactor with a single zone of Dirac-type catalysts for bimolecular Langmuir–Hinshelwood and mth-order reactions. Due to the strongly nonlinear solid–fluid interactions, a large variety of bifurcation diagrams was found in this case.

3.2 Multiple Reactions

3.2.1 Isothermal Conditions

In the previous section, reactor performance was optimized for single reactions using the *local* OCD approach. This implies that the activity distribution in each particle along the reactor is optimized independently, so as to promote locally the best attainable performance. In the case of multiple reactions this approach may not be adequate.

This has been demonstrated for an isothermal parallel reacting system (Wu, 1994)

$$\begin{aligned} A_1 &\rightarrow A_2 \\ A_1 &\rightarrow A_3 \end{aligned} \tag{3.26}$$

where A_2 is the desired product. The kinetic expressions considered for the two reactions are

$$r_1 = \frac{k_1 C_1^{m_1}}{(1 + K_1 C_1)^{b_1}} \tag{3.27a}$$

$$r_2 = \frac{k_2 C_1^{m_2}}{(1 + K_1 C_1)^{b_2}}. \tag{3.27b}$$

The dimensionless reactor mass balances are

$$\frac{dg_i}{dy} = \sum_{j=1}^{2} v_{ij} \, \mathrm{Da}_j \, (n+1) \int_0^1 f_j(\mathbf{u}) \, a(s, y) s^n \, ds, \qquad i = 1, \ldots, 3 \qquad (3.28)$$

which for a Dirac catalyst distribution become

$$\frac{dg_i}{dy} = \sum_{j=1}^{2} v_{ij} \, \mathrm{Da}_j \, f_j(\bar{\mathbf{u}}). \qquad (3.29)$$

Since the reaction rates (3.27) depend only on the concentration of a single component, the vector $\mathbf{u}$ is in reality simply u_1.

The reactor selectivity and yield are defined by

$$S = \frac{g_2(1)}{1 - g_1(1)} \qquad (3.30)$$

$$Y = g_2(1) = XS \qquad (3.31)$$

where X is the reactor conversion. According to the local OCD approach, the optimal value of $\bar{u}_1$ is determined as the one which locally maximizes the desired reactor performance. In the case where such a value of $\bar{u}_1$ is larger than the corresponding concentration value in the bulk, g_1, the optimal performance is not obtainable and we have the suboptimal situation where $\bar{u}_1 = g_1$ and $\bar{s} = 1$. It is in this event that the *global* OCD approach can provide better performance. In this case the $\bar{u}_1$ value is obtained numerically by a one-parameter optimization procedure where now the objective function is not the local but rather the global reactor performance. In Figure 3.6, with reference to the reacting system above, it is shown that the global OCD approach can result in higher reactor yield than the local approach. A difference in the performance index obtained by the two methods exists only when the reactor has two sections: one with $\bar{s}_{\mathrm{opt}} < 1 \, (0 < y < \hat{y})$ and the second with $\bar{s}_{\mathrm{opt}} = 1 (\hat{y} < y < 1)$. The difference increases when the length of the second section $(1 - \hat{y})$ increases. Similarly, the global OCD approach can also give higher reactor selectivity than the local OCD approach (Wu, 1994).

An alternative design approach, which utilizes a one-zone reactor of Dirac catalysts but still results in the same optimum performance obtained by the global OCD approach, has been proposed (Sheintuch et al., 1986; Wu, 1995). This was accomplished by optimizing the feed distribution along the reactor. In particular, Wu (1995) considered the case of two isothermal parallel reactions following Langmuir–Hinshelwood kinetics (3.27), with yield as the performance index. Utilizing the maximum principle, the optimal feed distribution was found analytically to be a uniform distribution of the feed stream along a reactor portion starting from its entrance, so the optimization problem was reduced to determining the length of this portion and the optimal catalyst location. For low Damköhler values, the feed distribution portion was the whole reactor length and the associated optimal catalyst location was a subsurface position. For large Damköhler values, the feed distribution portion was shorter than the reactor length and the optimal catalyst location was the pellet external surface.

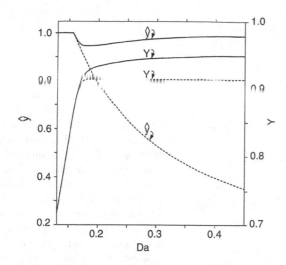

Figure 3.6. Maximum reactor outlet yield Y and reactor portion with subsurface optimal catalyst location, γ, as a function of Damköhler number Da. Reaction system given by equations (3.27), $m_1 = 1$, $b_1 = 2$, $m_2 = 1$, $b_2 = 1$, $\sigma = 20$, $n = 2$, $\phi_1^2 = 0.2$, $\kappa = 1$. Global OCD approach (solid curves), local OCD approach (dashed curves). (From Wu, 1994.)

A comparison of the optimal feed distribution (OFD) and the global OCD approaches is shown in Figure 3.7. Both approaches lead to the same outlet yield, and the length of the feed distribution for the OFD approach is equal to the length of the reactor portion with subsurface optimal catalyst location for the OCD approach. These similarities are not coincidental and are directly related to the fact that the $\bar{u}_1$ profile along the reactor is the same. For the OCD approach this profile is achieved by adjusting the catalyst location in pellets at each point along the reactor; for the OFD approach it is achieved by properly distributing the feed. Therefore the two approaches give the same reactor performance and differ only in the way they achieve it.

3.2.2 Nonisothermal Conditions

One- and two-zone reactor optimization has been reported for the case of ethylene epoxidation on silver catalyst taking place in a nonisothermal, nonadiabatic fixed-bed reactor (Baratti et al., 1994). The reaction scheme considered is

$$C_2H_4 + \tfrac{1}{2}O_2 \xrightarrow{1} C_2H_4O$$

$$C_2H_4 + 3O_2 \xrightarrow{2} 2CO_2 + 2H_2O$$

(3.32)

which follows the reaction kinetics (Klugherz and Harriott, 1971)

$$r_1 = \frac{k_1 C_1 C_3^2}{(0.0106 + 2144C_1 + 805C_3)(1 + 1271\sqrt{C_3})}$$

(3.33a)

$$r_2 = \frac{k_2 C_1 C_3^2}{(0.008 + 4166C_1 + 1578C_3)(1 + 718\sqrt{C_3})}$$

(3.33b)

where C_1 and C_3 are the concentrations (mol/cm^3) of ethylene and oxygen, respectively. This is a parallel reaction scheme similar to (3.26). Yield and selectivity

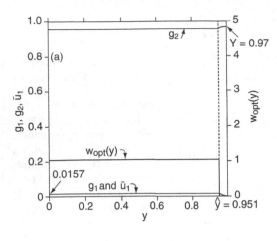

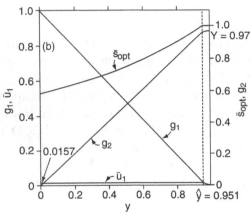

Figure 3.7. (a) Optimal dimensionless feed distribution along the reactor, $w_{opt}(y)$, corresponding dimensionless fluid-phase concentrations g_1 and g_2, and dimensionless concentration at the catalyst location, $\bar{u}_1$. The catalyst location is at the external surface of pellets. Outlet yield maximized using the OFD approach. (b) Optimal catalyst location $\bar{s}_{opt}$, corresponding dimensionless fluid-phase concentrations g_1 and g_2, and dimensionless concentration at the catalyst location, $\bar{u}_1$. Outlet yield maximized using the global OCD approach. Reaction system given by equations (3.27), $m_1 = 1$, $b_1 = 2$, $m_2 = 1$, $b_2 = 1$, $\sigma = 20$, $n = 2$, $\phi_1^2 = 0.2$, $\kappa = 0.5$, Da $= 0.25$. (From Wu, 1995.)

used as the performance indexes are given by equations (3.30) and (3.31), where g_1 and g_2 are the dimensionless bulk fluid concentrations of ethylene and ethylene oxide, respectively. The fluid-phase mass and energy balances for a one-zone reactor are

$$\frac{dg_i}{dy} = \sum_{j=1}^{2} v_{ij}\,\mathrm{Da}_j\,(n+1)\int_0^1 f_j(\mathbf{u}, \theta)\,a(s)s^n\,ds \tag{3.34}$$

$$\frac{d\tau}{dy} = \sum_{j=1}^{2} \bar{\beta}_j\,\mathrm{Da}_j\,(n+1)\int_0^1 f_j(\mathbf{u}, \theta)\,a(s)s^n\,ds - \mathrm{St}(\tau - \tau_c) \tag{3.35}$$

which for a Dirac catalyst distribution become

$$\frac{dg_i}{dy} = \sum_{j=1}^{2} v_{ij}\,\mathrm{Da}_j\,f_j(\bar{\mathbf{u}}, \bar{\theta}) \tag{3.36}$$

$$\frac{d\tau}{dy} = \sum_{j=1}^{2} \bar{\beta}_j\,\mathrm{Da}_j\,f_j(\bar{\mathbf{u}}, \bar{\theta}) - \mathrm{St}(\tau - \tau_c) \tag{3.37}$$

where St is the Stanton number. The solid-phase mass and energy balances (2.94)–(2.96) were reduced to algebraic equations using standard techniques. Optimization was carried out numerically, through one- or multiparameter estimation methods for the one- and two-zone reactors, respectively.

For small values of the Thiele modulus, the optimal catalyst location for yield maximization was mostly in the pellet interior. In this case, no discernible difference in the performance of the optimal, surface, and uniform distributions was observed, because kinetic control prevailed. Increasing catalyst activity led to ignition behavior for both surface and uniform catalysts. For the optimal Dirac distribution, though, intraparticle gradients present for subsurface catalysts helped to avoid reactor runaway. Thus, *when using Dirac catalysts it is possible to operate the reactor with higher loading (activity) catalysts, while avoiding reactor runaway.*

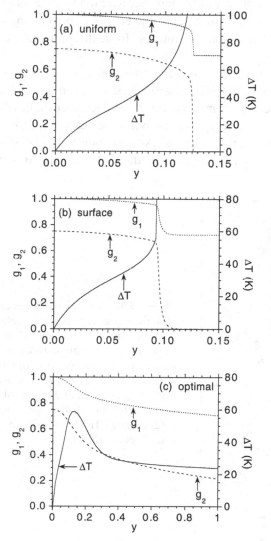

Figure 3.8. Dimensionless fluid-phase concentrations for ethylene (g_1) and oxygen (g_2) and temperature rise ΔT along the reactor for $\phi = 0.5$ for (a) uniform catalyst, (b) surface catalyst, (c) optimal catalyst for yield maximization ($\bar{s}_{opt} = 0.8$). (From Baratti et al., 1994.)

This has a beneficial effect on reactor performance, since while ignition results in low yields for the uniform and surface distributions, the yield obtained with the optimal Dirac catalyst continues to increase with catalyst activity. This behavior is clearly demonstrated in Figure 3.8, where dimensionless bulk-fluid concentrations and temperature rise (above the inlet reactor temperature) along the reactor are shown for the three distributions, using parameter values representative of typical reaction conditions. The surface pellet ignites earlier than the uniform, because there are no intraphase concentration gradients to slow down reaction rates. The runaway situation is not observed for the optimal Dirac distribution. In this case, there is a hot spot close to the reactor entrance, but eventually the temperature stabilizes.

For selectivity maximization only two positions were found to be optimal, depending primarily on the value of Thiele modulus: surface and center. This behavior is directly related to the complexity of the intrinsic reaction kinetics. However, subsurface locations were associated with negligible ethylene conversion, so that the surface distribution is the only one of interest from the practical point of view.

In Figure 3.9, a comparison between one- and two-zone reactors is shown for yield maximization. The optimal catalyst location in the second zone, $\bar{s}_{2,\text{opt}}$, is larger than in the first, $\bar{s}_{1,\text{opt}}$, in order to compensate for the decreased bulk reactant concentration. The optimal location $\bar{s}_{\text{opt}}$ for the one-zone reactor falls between $\bar{s}_{1,\text{opt}}$ and $\bar{s}_{2,\text{opt}}$, since it is in a sense their average. As the catalyst activity increases, the optimal location for the second zone reaches the pellet surface. It does not

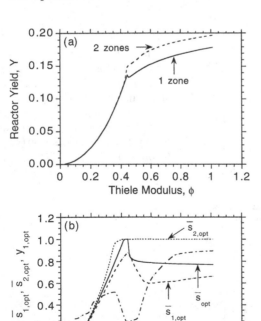

Figure 3.9. Comparison between one- and two-zone reactors for yield maximization in ethylene epoxidation. (a) Yield Y as a function of Thiele modulus ϕ. (b) Optimal catalyst locations for one-zone reactor ($\bar{s}_{\text{opt}}$) and for two-zone reactor ($\bar{s}_{1,\text{opt}}, \bar{s}_{2,\text{opt}}$) and dimensionless optimal length of the first zone, $y_{1,\text{opt}}$, as a function of Thiele modulus ϕ. (From Baratti et al., 1994.)

move to subsurface locations as $\bar{s}_{opt}$ and $\bar{s}_{1,opt}$ do, because in the second zone reactant concentrations are low, thus preventing runaway. The improvement by using a two-zone reactor is insignificant at low values of Thiele modulus, but at larger values yield improvements of about 15% over the single-zone reactor can be realized.

The fact that optimization of catalyst distribution leads to avoidance of reactor runaway in ethylene epoxidation is attractive. Since the desired reaction has lower activation energy than the undesired reaction, excessive temperatures result in low selectivity and yield. Hence, in order to maximize yield and selectivity, reactor runaway must be avoided. It is worth noting that axial temperature gradients can also be suppressed by using uniform pellets and diluting them appropriately with inert pellets or by using uniform pellets but with catalyst loading depending on their position along the reactor. In this way, a nonuniform activity profile is established along the reactor length. Both approaches have been shown to prevent reactor runaway and can also lead to improved selectivity (Caldwell and Calderbank, 1969; Narsimhan, 1976; Sadhukhan and Petersen, 1976; Pirkle and Wachs, 1987).

3.3 Experimental Studies

A few experimental studies describing the effects of nonuniform catalyst distributions in fixed-bed reactors have been reported in the literature. They are primarily concerned with oxidation reactions and are described below. Additional related studies in the context of automotive exhaust catalysis are reported in section 6.1.

3.3.1 Propane and CO Oxidation

Propane oxidation was investigated on Pt/γ-Al_2O_3 catalyst by Kunimori et al. (1982) in a fixed-bed reactor under isothermal conditions. They found that the observed activity was as follows: eggshell > uniform > egg-white, which is expected for first-order kinetics. However, in poisoning experiments, the egg-white catalyst was more resistant, which is consistent with a pore-mouth poisoning mechanism. These investigators also studied CO oxidation on the same catalyst. As noted earlier, this reaction exhibits bimolecular Langmuir–Hinshelwood kinetics, and egg-white catalysts can be more active. Indeed, the activity was in the order egg-white > uniform > eggshell. Examining the steady-state multiplicity behavior of the above system, it was found that the catalyst distribution alters the hysteresis behavior (Kunimori et al., 1986). The diffusional resistance introduced by the outer inert layer of the egg-white catalyst increases the width of the conversion-vs-temperature hysteresis loop. It is interesting to note that no difference in activity was observed between the powdered catalysts obtained by pulverizing the egg-white and eggshell catalyst pellets, which indicates that the preparation procedure did not alter the intrinsic catalyst activity.

Lee and Varma (1988) studied experimentally the behavior of an isothermal fixed-bed reactor for CO oxidation. Four different Pt/γ-Al_2O_3 catalysts with step activity distribution were considered. The steps were very thin (not larger than 8%

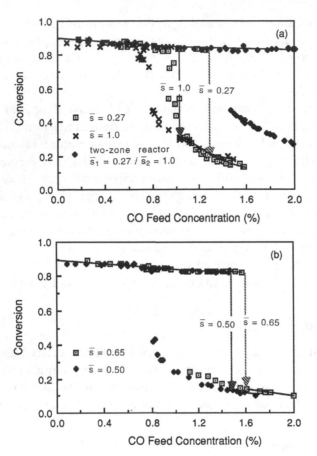

Figure 3.10. CO oxidation on Pt/γ-Al$_2$O$_3$ catalyst at 115°C and 152-ml/min flow rate. Comparison of one-zone reactors: (a) $\bar{s} = 0.27$ and $\bar{s} = 1.0$, (b) $\bar{s} = 0.65$ and $\bar{s} = 0.50$. In (a) results for the two-zone reactor ($\bar{s}_1 = 0.27$, $\bar{s}_2 = 1.0$) are also shown. (From Lee and Varma, 1988.)

of the pellet radius) and located approximately at $\bar{s} = 0.27, 0.50, 0.65, 1$. Two steady states were observed in all single-zone reactors. As shown in Figure 3.10, among the four types of catalysts studied, the reactor packed with the catalyst where the active layer was positioned at $\bar{s} = 0.65$ had the widest range of high-conversion regime. This range was further increased by using a dual-zone reactor, with $\bar{s} = 0.27$ and $\bar{s} = 1$ in the first and second zones, respectively. This improvement was expected according to the discussion in section 3.1.1. A direct experimental test of the effect of nonuniform catalyst distribution was demonstrated in a dual-zone reactor. Using the same catalyst pellets, the reactor performance (including the number of steady states) was dramatically altered when the packing order was reversed, i.e., first $\bar{s} = 0.27$ and second $\bar{s} = 1$ in the two zones, or vice versa, as shown in Figure 3.11. All of the above observations were in excellent qualitative agreement with theoretical predictions.

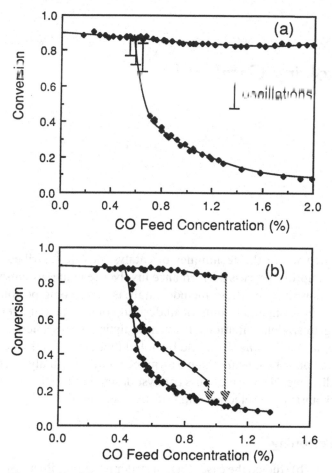

Figure 3.11. CO oxidation on Pt/γ-Al$_2$O$_3$ catalyst at 100°C and 152-ml/min flow rate. Comparison of two-zone reactors: (a) $\bar{s}_1 = 0.27$ and $\bar{s}_2 = 1.0$, (b) $\bar{s}_1 = 1.0$ and $\bar{s}_2 = 0.27$. (From Lee and Varma, 1988.)

3.3.2 Catalytic Incineration of Volatile Organic Compounds

Incineration of Volatile Organic Compounds was investigated by Frost et al. (1990) in a nonisothermal fixed-bed reactor with axial mass and heat dispersion and both inter- and intraparticle mass and heat transfer resistances, loaded with Pt/γ-Al$_2$O$_3$ pellets where the platinum was distributed as a thin layer slightly below the external surface. The reaction was assumed to follow Langmuir–Hinshelwood kinetics. The authors developed a model, which was fitted to experimental data on conversion vs inlet temperature by optimizing the values of the kinetic parameters and the effective diffusivity of reactants in the pellet. The model was used to size commercial-scale catalytic incinerators.

4

Studies Involving Catalyst Deactivation

A n important area where the techniques of catalyst design described above can be used to improve process performance involves reacting systems whose activity changes with time. These include catalysts undergoing poisoning as well as processes where small amounts of undesired components are removed from a stream (e.g. hydrodemetallation). In such catalytic systems a new variable is introduced: the *catalyst lifetime*. It should be noted that only the initial catalyst distribution can be controlled, while the performance index is usually integrated over the catalyst lifetime. Next, we discuss catalyst design in the presence of two deactivation mechanisms: nonselective and selective poisoning.

4.1 Nonselective Poisoning

Becker and Wei (1977b) studied the case of a first-order reaction with nonselective poisoning. They assumed that the poison adsorbs on the entire catalyst surface with no preference for the active element or the support (nonselective poisoning), and without ever reaching saturation. The diffusion–reaction process for the poison is governed by the poison Thiele modulus defined as

$$\phi_p = R\sqrt{k_p/D_{e,p}} \tag{4.1}$$

which represents the ratio between the rate of poison deposition (assumed to be first-order with respect to poison concentration) and the rate of poison diffusion inside the pellet. In this work, all diffusion coefficients were assumed to be constant with respect to time, which is true as long as the poison loading is not sufficiently large to significantly reduce the catalyst pore-mouth. Depending on the value of the poison Thiele modulus, two characteristic cases are obtained. For $\phi_p \ll 1$, the poison deposits uniformly throughout the pellet (*uniform poisoning*), while for $\phi_p \gg 1$ the poison deposits preferentially in a thin layer close to the pellet's external surface (*pore-mouth poisoning*). Four types of active element distributions were considered: uniform, eggshell, egg-white, and egg-yolk. When the main reaction is under diffusion control (large ϕ_0), as in the nondeactivating case (section 2.1.1), the best performance for uniform poisoning is provided by

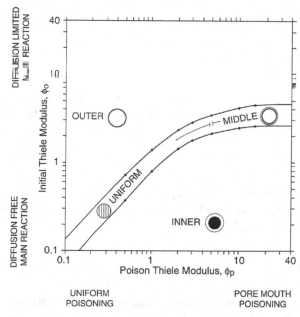

Figure 4.1. Catalyst selection chart for maximization of catalyst lifetime. (Active-catalyst volume)/(total support volume) = 1/3. First-order reaction with first-order poison deposition. (From Becker and Wei, 1977b.)

an eggshell distribution. However, under pore-mouth poisoning conditions, a significant improvement in catalyst performance can be achieved by distributing the active element in the inner region of the pellet. These considerations lead to the catalyst selection chart shown in Figure 4.1.

Similar improvement under pore-mouth poisoning was also demonstrated for uniform catalysts covered with an inert protective layer composed of a different support material, hence with different diffusion properties (Wolf, 1977; Polomski and Wolf, 1978). For CO oxidation, this type of catalyst showed an enhancement in activity when operating in the region of high reactant concentrations where the reaction exhibits negative-order kinetics, whereas at lower concentrations, where the reaction is essentially first-order, only the catalyst lifetime could be improved.

Bacaros et al. (1987) and Pavlou and Vayenas (1990b) addressed the problem of optimal initial activity distribution in isothermal pellets undergoing pore-mouth poisoning. First-order kinetics were examined, and it was assumed that catalyst poisoning was much slower than diffusion within the pellet, so that the quasi–steady-state assumption could be employed in the mass balance of the main reactant. The dimensionless model equations are

$$\frac{1}{s^n}\frac{d}{ds}\left(s^n\frac{du}{ds}\right) = a(s)\phi^2 u, \qquad 0 < s < s_p \tag{4.2a}$$

$$\frac{1}{s^n}\frac{d}{ds}\left(s^n\frac{du}{ds}\right) = 0, \qquad s_p < s < 1 \tag{4.2b}$$

$$s = 0: \qquad \frac{du}{ds} = 0 \tag{4.3a}$$

$$s = 1: \qquad u = 1 \tag{4.3b}$$

where s_p corresponds to the location of the deactivation front. Catalyst poisoning was described by the Voorhies equation, which implies that the front s_p advances towards the pellet interior independently of the concentrations of the catalyst, reactants, or poison:

$$s_p = 1 - \Theta^\pi \tag{4.4}$$

where Θ is dimensionless time and π is an empirical coefficient ranging from 0.5 to 1 (Carberry, 1976). The performance index used for optimization was the average reactant conversion over the catalyst lifetime,

$$\mathcal{H} = \int_0^1 \eta(\Theta)\, d\Theta. \tag{4.5}$$

At low Thiele modulus ($\phi \rightarrow 0$), it was found that the optimal distribution is a Dirac-delta located at the pellet center. This is because under kinetic control, any catalyst distribution provides the same performance if no deactivation occurs, while the time required for the poison to reach the catalyst is maximum if the catalyst is located at the pellet center.

For large Thiele-modulus, the optimal initial activity distribution tends to be bimodal, concentrating one portion of the active element near the pellet's external surface and another portion close to its center (Pavlou and Vayenas, 1990b), as illustrated in Figure 4.2.a for $\pi = 1$. The specific shape depends on the pellet geometry, values of the physicochemical parameters involved, and kinetics of the deactivating process through the value of the parameter π. The catalyst loading required near the center increases in going from the slab to the sphere geometry,

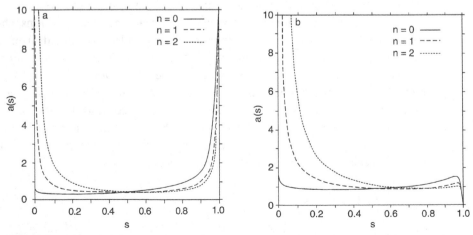

Figure 4.2. Optimal catalyst distribution for nonselective poisoning. First-order main reaction with (a) constant velocity of the deactivating front ($\pi = 1$), (b) diffusion-controlled deactivation ($\pi = 0.5$). Here $\phi[(n+1)/g_v]^{1/2} = 20$. (From Pavlou and Vayenas, 1990b.)

with a concomitant decrease near the external surface. When $\pi = 0.5$, the catalyst loading is substantially smaller near the surface and in fact vanishes at the surface, as shown in Figure 4.2.b. This behavior results from the interaction of two opposing factors. One is the effectiveness factor, which dictates that the catalyst should be placed close to the surface. The other is the catalyst lifetime, which dictates that it should be placed near the center. As seen in Figure 4.2.a, these factors result in a bimodal optimal distribution for $\pi = 1$. However, when $\pi = 0.5$, the poisoning front moves faster at the beginning. Consequently, the optimal catalyst loading near the pellet surface decreases substantially. Finally, it should be noted that for the case examined, contrary to nondeactivating systems, depositing the catalyst nonuniformly does not improve the pellet performance significantly.

For an isothermal first-order reaction, the case of uniform poisoning, in both single pellets and fixed-bed reactors, was considered by DeLancey (1973). The objective function included economic considerations related to conversion of reactants and catalyst replacement. An approximate solution was derived using Pontryagin's principle. For a single pellet the optimal catalyst distribution was eggshell, while for a fixed-bed reactor it was uniform.

4.2 Selective Poisoning

In contrast to the case of nonselective poisoning discussed above, where only specific situations have been investigated, a result of general validity has been obtained in the case of selective poisoning (Brunovska et al., 1990). In general, the poisoning rate depends on the concentration of poison species, the concentrations of reactants and products, and the temperature. Since the poisoning rate is usually much lower than that of the main reaction, the quasi–steady-state approximation can be adopted. In addition, it was assumed that the rate of deactivation depends linearly on the local active catalyst concentration. The model equations in dimensionless form are

$$\frac{1}{s^n} \frac{d}{ds} \left(s^n \frac{du}{ds} \right) = a(s,t) \phi^2 f(u, u_p, \theta) \tag{4.6}$$

$$\frac{1}{s^n} \frac{d}{ds} \left(s^n \frac{du_p}{ds} \right) = a(s,t) \phi_p^2 f_p(u, u_p, \theta) \tag{4.7}$$

$$\frac{1}{s^n} \frac{d}{ds} \left(s^n \frac{d\theta}{ds} \right) = -a(s,t) \beta \phi^2 f(u, u_p, \theta) \tag{4.8}$$

with BCs

$$s = 0: \qquad \frac{du}{ds} = 0, \quad \frac{du_p}{ds} = 0, \quad \frac{d\theta}{ds} = 0 \tag{4.9a}$$

$$s = 1: \qquad u = 1, \quad u_p = 1, \quad \theta = 1. \tag{4.9b}$$

The deactivation is accounted for by a balance of active sites, which in terms of the activity distribution function takes the form

$$\frac{\partial a}{\partial \Theta} = -a f_p(u, u_p, \theta) \tag{4.10}$$

with initial condition

$$\Theta = 0: \qquad a = a(s, 0) \tag{4.11}$$

An economic criterion based on profit per unit time was used as the performance index, which takes into account price of product and cost of catalyst replacement:

$$\mathcal{G} = \frac{\alpha_1 \int_0^{\Theta_{oper}} \eta(\Theta) \, d\Theta - \alpha_3}{\Theta_{oper}} \tag{4.12}$$

where Θ_{oper} is the dimensionless operating time and α_1, α_3 are weighting coefficients proportional to product value and catalyst cost, respectively. The optimization variables were the initial catalyst distribution $a(s, 0)$ and the operating time. By developing an optimality condition and following similar arguments to those in section 2.3.2, it was shown that the optimal distribution is a Dirac-delta function if no upper bound is imposed on the local catalyst loading. In the case where local loading is bounded, the optimal distribution is expected to take the form of a step function, in analogy with nondeactivating systems described in section 2.5.3. This result is valid for multiple reactions, following arbitrary kinetics, in the presence of thermal effects as well as inter- and intraparticle mass and heat transport resistances.

The case of an isothermal first-order reaction where the poisoning kinetics is also first-order was investigated as an example (Brunovska et al., 1990). It was found that for low values of the ratio of the two Thiele moduli characteristic of the poisoning and main reactions, ϕ_p/ϕ, the optimal location is at the pellet external surface, because the intraparticle resistance is larger for the reactant than for the poison. For increasing values of this parameter, the optimal location moves towards the pellet interior, to an extent which depends on the specific operating conditions. Further, the optimal operating time increases monotonically with ϕ_p/ϕ.

The case of selective poisoning in a fixed-bed reactor was studied by Markos and Brunovska (1989) for a nonisothermal plug-flow reactor. Optimization was carried out only for Dirac-type distributions constituting a single-zone reactor, and in all cases significant improvements were reported when using the appropriate Dirac-delta distribution. First-order deactivation kinetics (on poison concentration) and first-order kinetics of the main reaction (on reactant concentration) were considered. The objective function took account of the catalyst and raw-materials cost as well as the value of the products. It was defined similarly to (4.12), but in this case the integral of outlet reactor conversion was used

$$\mathcal{K} = \frac{(\alpha_1 - \alpha_2) \int_0^{\Theta_{oper}} X \, d\Theta - \alpha_3}{\Theta_{oper}} \tag{4.13}$$

where α_2 is a weighting coefficient proportional to the cost of raw materials.

Two competing factors determine the optimal catalyst location. Surface locations favor high initial reactor conversion, while internal locations are beneficial for longer catalyst lifetime. This can be seen in Figure 4.3, where the objective function is shown as a function of the operating time Θ_{oper} for various active

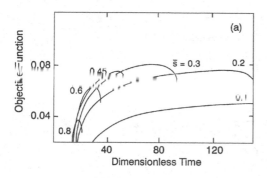

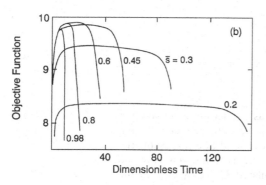

Figure 4.3. Time dependence of objective function given by equation (4.13) for various active-layer locations; $\phi_p = 10$, $Da_p = 10$, (a) $(\alpha_1 - \alpha_2)/\alpha_3 = 0.1$, (b) $(\alpha_1 - \alpha_2)/\alpha_3 = 10$. Isothermal first-order reaction, poisoning kinetics first-order. (From Markos and Brunovska, 1989.)

catalyst locations, and cases of an expensive [i.e. $\alpha_3 \gg (\alpha_1 - \alpha_2)$] or cheap [i.e. $\alpha_3 \ll (\alpha_1 - \alpha_2)$] catalyst. In the first case (Figure 4.3.a), the objective function is maximized for $\bar{s} = 0.3$ and $\Theta_{oper} \sim 80$, while in the second (Figure 4.3.b), the optimal performance is reached for $\bar{s} \sim 0.6$ and $\Theta_{oper} \sim 20$. Thus, when the catalyst cost is high, it is better to locate the active element in the pellet interior and have long operation time, while if the catalyst cost is low, the optimal location moves towards the pellet surface in order to favor the main reactant transport process, and the operating time decreases. The case of a single second-order reaction and of two first-order consecutive reactions was also investigated numerically by Markos et al. (1990), using an isothermal plug-flow heterogeneous reactor model.

4.3 Experimental Studies

In this section, we present results from a variety of experimental studies related to the behavior of nonuniform catalyst distributions in the presence of deactivation. These include methanation, olefin hydrogenation, and NO reduction reactions. Catalyst design under poisoning conditions is critical in the context of automotive exhaust catalysis, and is discussed in section 6.1.

4.3.1 Methanation

The methanation reaction

$$CO + 3H_2 \rightarrow CH_4 + H_2O \tag{4.14}$$

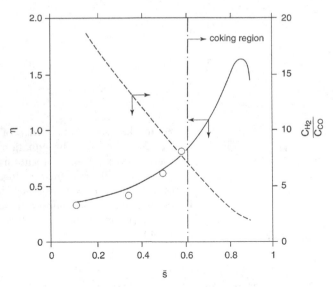

Figure 4.4. Effectiveness factor as a function of active-layer location for step-type Ni/Al_2O_3 pellets during CO methanation in the presence of coking. The solid line represents the simulation results, the circles denote experimental data, and the dashed line shows the calculated C_{H_2}/C_{CO} ratio at the active layer location. $T = 573$ K, $C_{CO,f} = 5 \times 10^{-6}$ mol/cm^3, $C_{H_2,f} = 10.5 \times 10^{-6}$ mol/cm^3. (From Wu et al., 1990b.)

over Ni/Al_2O_3 was studied in step-type catalyst pellets by Wu et al. (1990b). This reaction is accompanied by selective poisoning due to carbon deposition (coking) when the H_2/CO ratio is low. The pellets were prepared by pressing an active powder (Ni/Al_2O_3) layer sandwiched between two inert powder (Al_2O_3) layers. It was found that significant coking occurred only when the active layer was located close to the external surface of the pellet. By placing the catalyst layer deeper inside, diffusional resistances introduced by the outer Al_2O_3 layer increased the H_2/CO ratio at the catalyst location, since the diffusivity of hydrogen is larger than that of carbon monoxide in the Knudsen diffusion regime. High H_2/CO_4 ratios suppress coking, but if the catalyst is placed too deep inside the pellet, the effectiveness factor decreases significantly. Accordingly, for the reaction conditions investigated, a pellet with the catalyst layer centered at $\bar{s} \sim 0.6$ had the highest effectiveness factor while simultaneously avoiding coking, as shown in Figure 4.4. Agreement of experiments with theoretical diffusion–reaction calculations was good, as also shown in the figure.

4.3.2   Hydrogenation

Ethylene hydrogenation in the presence of low concentrations of the poison thiophene on Pt/Al_2O_3 catalysts was studied by Brunovska and coworkers (Pranda and Brunovska, 1993; Remiarova et al., 1993; Brunovska et al., 1994). Pellets were prepared both by impregnation of preformed support and by pressing active and inert powder layers. The performance index used for comparison was the total reactant conversion over the catalyst lifetime, equation (4.5).

Experiments demonstrated that the selective poisoning process consists of two consecutive steps: reversible adsorption of poison precursor, followed by slow irreversible surface reaction. Initial effectiveness factors were found to be larger for the surface catalyst pellets, indicating that mass transfer resistances decreased the hydrogenation rate. The maximum value of the performance index was found to depend on the operating time. For short operating times, the surface catalyst pellet showed slightly better performance, since deactivation was limited. For longer times subsurface distributions were optimal, showing that mass transfer resistance retarded pellet deactivation. In addition, the dependence of the performance index on catalyst location was not monotonic. In some cases two maxima appeared, whose location and relative magnitude depended on the operating time and width of the active layer. This behavior is due to the interaction between complex main and poisoning reaction kinetics. While maintaining the center of the active layer at a fixed subsurface location, the effect of layer width was also investigated. As shown in Figure 4.5, the thinnest layer (i.e. approaching the Dirac-delta distribution) exhibited the best performance, as expected from the theoretical results presented in section 4.2.

Lin and Chou (1994) investigated the selective hydrogenation of isoprene on $Pd/\gamma\text{-}Al_2O_3$ catalysts, which is representative of partial hydrogenation of pyrolysis gasoline from naptha cracking to increase its octane number. This is a consecutive reaction system with isoprene hydrogenating to isopentenes, which in turn hydrogenate to isopentane. Coke buildup from the polymerization of dialkenes is the main mechanism for catalyst deactivation. Eggshell and uniform catalysts were prepared by impregnation of the support with a toluene solution of $Pd(CH_3COO)_2$ and an aqueous solution of $Pd(NH_3)_4(NO_3)_2$, respectively. Both catalysts contained 0.2 wt% Pd with the same particle size of 8 Å. The selectivity to isopentenes was higher for the eggshell catalyst, since they diffused away from the reaction zone before hydrogenation to isopentane. The eggshell catalyst

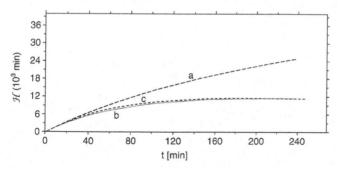

Figure 4.5. Performance index $\mathcal{H}$ given by equation (4.5) as a function of time for three step catalyst distributions: curve a, $\bar{s}_1 = 0.90$, $\bar{s}_2 = 0.93$; curve b, $\bar{s}_1 = 0.86$, $\bar{s}_2 = 0.97$; curve c, $\bar{s}_1 = 0.83$, $\bar{s}_2 = 1.0$. (From Pranda and Brunovska, 1993.)

also showed smaller coke formation and slower deactivation. This was attributed to the higher hydrogen partial pressure in the reaction zone, which inhibits coke buildup and further polymerization. In addition, coke precursors could diffuse out from the catalyst pores more easily, hence decreasing coke formation. However, palladium migration from the outside towards the core of the pellet occurred for eggshell catalysts (Lin and Chou, 1995), during hydrogenation as well as subsequent catalyst regeneration. This migration was accompanied by sintering, which resulted in loss of catalytic surface area.

4.3.3   NO Reduction

Stenger et al. (1988) and Hepburn et al., (1991b) studied the design of Rh/Al_2O_3 honeycomb catalysts for NO reduction with H_2, in the presence of SO_2, for use in stationary pollution sources such as power plants and industrial boilers. Eggshell and egg-white catalysts were prepared by coimpregnation of the support with $RhCl_3$ and HF. Electron probe microanalysis revealed that sulfur was distributed throughout the support, indicating that SO_2 was not limited by diffusion. Thus, subsurface location of Rh does not delay deactivation. It is worth noting that sulfur deposition on alumina was small, while higher sulfur concentrations were observed where Rh was present, indicating selective poisoning. The egg-white catalyst exhibited greater tolerance to deactivation than the eggshell catalyst, especially at longer periods. This behavior was attributed to the presence of the outer alumina layer, which scavenges the SO_2 before it reaches the noble metal.

5

Membrane Reactors

Membrane reactors offer the advantage over conventional fixed-bed reactors of combining chemical reaction and separation in a single unit. They can substantially improve the performance of reactions by selective removal of one of the reaction products or by controlled addition of a reactant. In the former case, conversion enhancement beyond the thermodynamic limit can be achieved for equilibrium-limited reactions, while in the latter, product selectivity can be improved by influencing the concentration and residence time of components giving rise to undesired reactions. Several articles which review experimental and theoretical studies of catalytic membrane reactors are available in the literature (Hsieh, 1991; Tsotsis et al., 1993; Saracco and Speccia, 1994; Zaman and Chakma, 1994).

5.1 Membrane Reactors with Nonuniform Catalyst Distribution

In this section we discuss theoretical studies addressing nonuniform distribution of catalyst in membrane reactors, in order to gain insight into the effect of catalyst location on this new reactor type, before addressing optimization issues in the following sections. Yeung et al. (1994) investigated the influence of the location of a Dirac-delta catalyst in pellets contained in an inert membrane reactor with catalyst on the feed side (IMRCF), a catalytic membrane reactor (CMR), and a conventional fixed-bed reactor (FBR). For the IMRCF and the FBR, the active catalyst is distributed in pellets placed inside the membrane or the reactor tube respectively, while for the CMR the active catalyst is distributed within the membrane itself as shown in Figure 5.1.

In order to provide analytical expressions so that the effects of the use of a membrane as well as nonuniform catalyst distribution are readily identified, an isothermal, first-order reversible reaction $A \rightleftharpoons B$ was considered, under well-mixed fluid-phase conditions and with no interphase mass transfer resistances. Furthermore, pure inert and pure A are fed to the permeate and feed sides, respectively of the IMRCF and CMR. The function of the inert is to sweep component B produced in the reactor and diffusing towards the permeate side. The steady-state

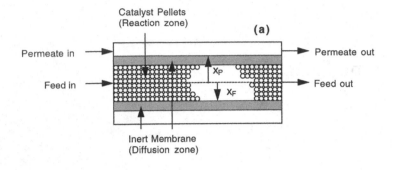

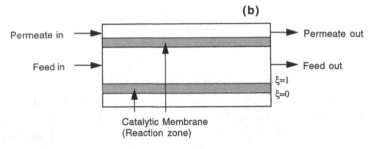

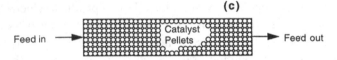

Figure 5.1. Schematic diagrams of (a) inert membrane reactor with catalyst on the feed side (IMRCF), (b) catalytic membrane reactor (CMR), and (c) fixed-bed reactor (FBR). (From Yeung et al., 1994.)

mass balances for the three reactors were solved analytically for uniform and Dirac catalyst distributions (see Szegner, 1997 for details). For the membrane reactors, the total conversion X_T is computed by combining the feed and permeate sides. It is thus defined as the ratio between the total molar flow of reactant converted to product and the initial molar flow of reactant on the feed and permeate sides:

$$X_T = \frac{\zeta_{F,A}^0 + \zeta_{P,A}^0 \Gamma_P^0 - \zeta_{F,A}\Gamma_F - \zeta_{P,A}\Gamma_P}{\zeta_{F,A}^0 + \zeta_{P,A}^0 \Gamma_P^0} \tag{5.1}$$

where $\zeta_{F,A}$, $\zeta_{P,A}$, Γ_F, and Γ_P are the reactant mole fractions and dimensionless molar flow rates on the feed and permeate sides, respectively, with

$$\Gamma_F = \frac{Q_F}{Q_F^0}, \qquad \Gamma_P = \frac{Q_P}{Q_F^0}. \tag{5.2}$$

The dimensionless residence time Θ is defined as the ratio between residence time and the characteristic time for diffusion:

$$\Theta = \frac{V_r P_F / R_g T Q_F^0}{\delta_M^2 / D_{e,A}} = \frac{D_{e,A} V_r P_F}{R_g T Q_F^0 \delta_M^2}. \tag{5.3}$$

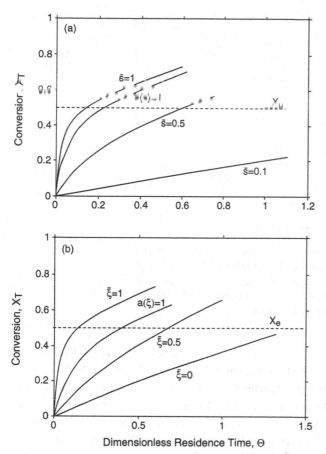

Figure 5.2. The effect of location of a Dirac-delta catalyst activity distribution on the total conversion X_T as a function of dimensionless residence time Θ for an isothermal first-order reversible reaction $A \rightleftharpoons B$, for well-mixed (a) IMRCF and (b) CMR. (From Yeung et al., 1994.)

The variation of the total conversion as a function of residence time is shown in Figure 5.2 for the two membrane reactors as the location of the Dirac catalyst is varied in the catalyst pellets (for the IMRCF) and in the catalytic membrane (for the CMR). In each case, the performance of the corresponding uniform catalyst distribution is also shown for comparison. For the IMRCF (Figure 5.2.a), it is clearly seen that on moving the catalyst from a position near the center ($\bar{s} = 0.1$) to the surface ($\bar{s} = 1$) of the pellets, the conversion X_T increases to larger values than in the uniformly distributed case [$a(s) = 1$]. This occurs because of decreased intraphase mass transfer resistance as the Dirac distribution is located closer to the pellet surface. A similar result is observed for the CMR (Figure 5.2.b) as the Dirac-delta distribution is moved from the permeate side ($\bar{\xi} = 0$) to the feed side of the membrane ($\bar{\xi} = 1$). When the catalyst is placed on the permeate side, the conversion does not exceed the equilibrium value, because in this case the

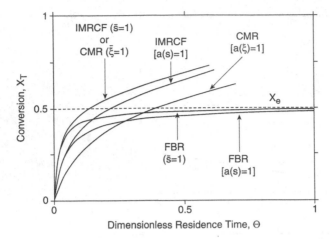

Figure 5.3. Total conversion X_T for well-mixed IMRCF, CMR, and FBR with uniform and Dirac-delta catalyst activity distributions, as a function of dimensionless residence time Θ, for an isothermal first-order reversible reaction $A \rightleftharpoons B$. (From Yeung et al., 1994.)

selective product separation does not play a role in enhancing the conversion. On comparing results of the two membranes (i.e. Figure 5.2.a and b) it is seen that the improvement in reactor performance on locating the Dirac catalyst at $\bar{s}$ or $\bar{\xi} = 1$, as compared to the uniform catalyst distribution, is greater for the CMR. This occurs because in the case of the uniform CMR, as the reactant diffuses through the membrane, the back reaction of the product competes with its selective removal. In the case of the IMRCF, the difference in the performance between the uniform and the surface distribution is not as large, because the separation occurs outside the reaction zone.

A comparison of the various reactor configurations is shown in Figure 5.3. The function of the membrane in this system is evident on considering that at large residence time the fixed-bed performance is limited to $X_T = 0.5$ by chemical equilibrium. That limit is exceeded by all membrane reactors. The Dirac-delta distribution located at the feed side or at the pellet surface (i.e. $\bar{\xi} = 1$ for the CMR and $\bar{s} = 1$ for the IMRCF and FBR) shows superior performance to a uniform activity profile. This distribution maximizes the reactant concentration at the catalyst site, since it is not limited by diffusion through the membrane or the pellet. The performance of the CMR and IMRCF is identical when the Dirac location is at the feed side of the membrane (for the CMR) or at the external surface of the pellet (for the IMRCF), due to the absence of mass transfer resistances. These two membrane configurations exceed the performance of the FBR configurations over the full range of residence time. However, for small Θ the FBR with $\bar{s} = 1$ (i.e. Dirac-delta at the external surface of the pellet) performs better than the IMRCF with $a(s) = 1$ (i.e. uniform catalyst distribution). This is because at small residence time, due to diffusional limitations, the reactant sees only a

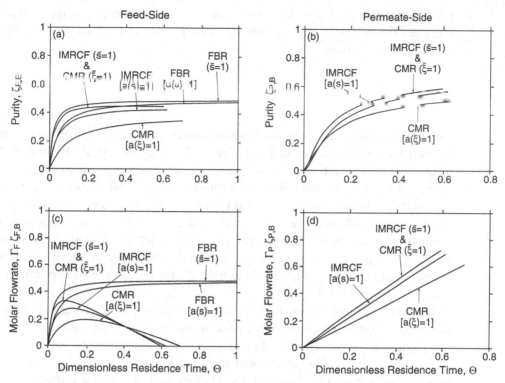

Figure 5.4. A comparison of well-mixed IMRCF, CMR, and FBR with uniform $[a(s) = 1$ or $a(\xi) = 1]$ and Dirac-delta (at $\bar{s} = 1$ or $\bar{\xi} = 1$) catalyst activity distributions for (a) product purity on the feed side $\zeta_{F,B}$, (b) product purity on the permeate side $\zeta_{P,B}$, (c) product molar flow rate on the feed side $\Gamma_F \zeta_{F,B}$, and (d) product molar flowrate on the permeate side $\Gamma_P \zeta_{P,B}$, as a function of dimensionless residence time Θ. The system is an isothermal first-order reversible reaction A $\rightleftharpoons$ B. (From Yeung et al., 1994.)

small amount of the total catalyst in the pellets of the IMRCF, but it is exposed to the full amount of catalyst in the FBR, thus resulting in higher conversion to product.

Similar arguments hold when product purity or product flow rate on either the feed or the permeate side is used as the performance index. The purities on the feed and permeate sides are the product mole fractions $\zeta_{F,B}$ and $\zeta_{P,B}$ respectively, while the product flow rates are $\Gamma_F \zeta_{F,B}$ and $\Gamma_P \zeta_{P,B}$. A comparison for these cases is illustrated in Figure 5.4. It can be seen (Figure 5.4.a) that on the feed side both FBRs perform better than the membrane reactors for most cases. This is because the feed side is depleted of the products as a result of diffusion across the membrane. This point is illustrated in Figure 5.4.c, where a maximum may be observed. For small residence times, the product molar flow rate is small because of insufficient contact with the catalyst, whereas for larger residence times the product permeates through the membrane until total permeation occurs (i.e. no flow out from the feed side). When the product purity or its flow rate on the permeate side is to be maximized, it is seen in Figure 5.4.b and d that for both the

CMR and IMRCF a Dirac-delta distribution of catalyst located on the feed side (i.e. $\bar{\xi} = 1$ for the CMR and $\bar{s} = 1$ for the IMRCF) shows better performance than a uniform distribution.

5.2   Optimal Catalyst Distribution in Pellets for an Inert Membrane Reactor

The problem of determining the optimal distribution of catalyst in pellets for an inert membrane reactor with catalyst on the feed side (IMRCF), with well-mixed fluid phases on both the feed and permeate sides, where multiple reactions with arbitrary kinetics take place under nonisothermal conditions and external transport resistances, has been solved (Yeung et al., 1994). The methodology is similar to that followed previously in section 2.5.1. In this case, in addition to the mass and energy balance equations for the pellet, fluid-phase balances for the feed and permeate sides are also required. The resulting set of equations is cumbersome and is presented along with the formulation of the optimization problem in Appendix B.

Due to the complexity and nonlinearity of the equations involved, the solution of the optimization problem can be challenging even using advanced numerical techniques. In the general case, direct numerical methods must be employed (cf. Baratti et al., 1993). However, when B equals zero (i.e. linear dependence of catalytic surface area on loading), the Hamiltonian becomes a linear function of the active element distribution $\mu(s)$, and then, as discussed in section 2.5.3, an analytical solution can be obtained. In this case, the optimal distribution is a multiple-step function, where the number of steps is controlled by the number of sign changes of the switching function. This means that the presence of an inert membrane and a permeate flow do not influence the *nature* of the optimal catalyst distribution, although they may well alter the specific locations and the number of steps. One would also expect that as the upper bound of local catalyst loading tends to infinity, the step distribution would approach a Dirac-delta distribution.

It should be clear that the similarity of the above optimization solution to that of a single pellet (section 2.5.3) exists because fluid phase conditions in the membrane reactor are well mixed. If the fluid phases are not well mixed, then one can expect that the optimal catalyst distribution in pellets will change with axial position in the reactor, as in the fixed-bed reactor in section 3.1. However, a rigorous proof of this conjecture is not yet available.

5.3   Optimal Catalyst Distribution in a Catalytic Membrane Reactor

The general problem of optimal catalyst distribution in a CMR has been studied by Szegner (1997). The problem formulation is similar to that for the IMRCF (see Appendix B), except that the catalyst is placed in the membrane itself and not in the pellets. Thus again, in general, a numerical approach is required to determine the optimal distribution that maximizes the Hamiltonian. In the case of $B = 0$, a multiple-step distribution is proven to be optimal.

The particular case of a planar catalytic membrane where the downstream bulk flow is very large has been considered by Keller et al. (1984). In this case we can assume that the downstream concentration of the reactant diffusing through the membrane is zero. Under isothermal conditions and in the absence of external transport resistances, the problem reduces to the diffusion–reaction equation in the catalytic membrane:

$$\frac{d^2 u}{ds^2} = \phi^2 a(s) f(u) \tag{5.4}$$

together with the boundary conditions

$$s = 0: \qquad u = 1 \tag{5.5a}$$
$$s = 1: \qquad u = 0. \tag{5.5b}$$

Note that this problem differs from that of a catalyst pellet, given by equations (2.63)–(2.66), only in the BCs. The space coordinate of the membrane in this case starts at the side in contact with the feed stream ($s = 0$) and finishes at the downstream side ($s = 1$).

For reaction rates $f(u)$ that increase monotonically with reactant concentration, and with effectiveness factor as the performance index, a Dirac-delta function located at the feed side surface of the membrane is found to be the optimal catalyst distribution, in accordance with the analysis in section 2.1. This is where the highest reactant concentration is encountered, which is beneficial for a positive-order reaction.

A different answer for the optimization problem is obtained when minimizing the flux of unreacted material leaving the membrane,

$$I = \left(-\frac{du}{ds} \right)_{s=1} . \tag{5.6}$$

Using a specific variational analysis applied to the case of linear kinetics, i.e. $f(u) = u$, Keller et al. (1984) found that the optimal activity distribution is given by the following *step function*:

$$a(s) = \begin{cases} 1/L, & s_2 \leq s \leq s_2 + L \\ 0, & 0 \leq s < s_2 \text{ and } s_2 + L < s \leq 1 \end{cases} \tag{5.7}$$

where

$$s_2 = 1 - L - \frac{\sqrt{L}}{\phi} \tag{5.8a}$$

$$L = 1 + 2 \left(\frac{1 - \sqrt{1 + \phi^2}}{\phi^2} \right) . \tag{5.8b}$$

Even though both the objective functional (5.6) and the BCs (5.5) are not the same as in the case of the catalyst pellet, it is surprising that here we encounter for the first time an optimal activity distribution which is not a Dirac-delta function. In order to understand this point, following the procedure described in section

2.3.1, we can derive the necessary condition for optimality,

$$\int_0^1 G^* a^*(s)\, ds \geq \int_0^1 G^* a(s)\, ds \tag{5.9}$$

which is the same as (2.81) with $n = 0$, except that the function G^* is now given by

$$G^* = \phi^2 \lambda u^* \tag{5.10}$$

and the Lagrange multiplier λ satisfies

$$\frac{d^2 \lambda}{ds^2} = \phi^2 a^* \lambda \tag{5.11}$$

with BCs

$$s = 0: \qquad \lambda = 0 \tag{5.12a}$$

$$s = 1: \qquad \lambda = 1. \tag{5.12b}$$

By taking (5.7) as the optimal distribution a^* and substituting in equations (5.10)–(5.12), we obtain

$$G^* = \begin{cases} \phi^2 s(1 - s - L)(1 - L)^{-2} \exp(-\phi\sqrt{L}), & 0 \leq s < s_2 \\ \phi^2 [1 - s_2(1 - L)^{-1}]^2 \exp(-\phi\sqrt{L}), & s_2 \leq s \leq s_2 + L . \\ \phi^2 (1 - s)(s - L)(1 - L)^{-2} \exp(-\phi\sqrt{L}), & s_2 + L < s \leq 1 \end{cases} \tag{5.13}$$

It can be seen that in the interval $s_2 \leq s \leq s_2 + L$, where the membrane is catalytically active (i.e. $a^* \neq 0$), the function $G^*(s)$ is constant and attains its maximum value. Therefore, we are in the peculiar situation discussed in the context of equation (2.89), where the necessary condition (5.9) cannot be violated by selecting an activity distribution $a(s)$ with the shape of a Dirac-delta function. This is in agreement with the result of Keller et al. (1984), who conclude that the optimal distribution is a step and not a Dirac-delta function.

5.4   Experimental Studies

Because the topic of optimal catalyst distribution in membrane reactors started receiving attention only recently, few experimental studies have been reported. The first set consists of dehydrogenation reactions in packed-bed membrane reactors, where the catalyst distribution has been shown to influence reactor performance. Secondly, preparation of catalytic membranes with controlled active layer thickness and location is addressed.

5.4.1   Dehydrogenation Reactions

Cannon and Hacskaylo (1992) demonstrated that catalyst location and amount in the membrane can affect reactor performance. They investigated cyclohexane dehydrogenation in porous Vycor glass tubes packed with 0.5-wt% Pd/Al_2O_3 catalyst pellets. Three membranes were used: (a) a pure Vycor glass membrane, (b) a

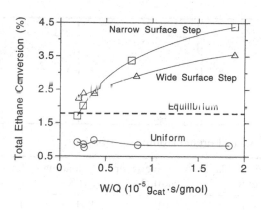

Figure 5.5. Effect of residence time, W/Q, on the conversion of ethane in an IMRCF with different catalyst distributions during ethane dehydrogenation on Pt-Sn/γ-Al$_2$O$_3$ pellets. Feed: 20% C$_2$H$_6$/Ar. (From Yeung et al., 1996.)

Vycor glass membrane containing 0.25% Pd, which was located preferentially at the inner and outer surfaces of the glass tube (PM1), and (c) a Vycor glass membrane containing 0.5% Pd, which was distributed uniformly throughout the membrane wall (PM2). In all cases, total cyclohexane conversion increased with increasing sweep-gas flow rate on the permeate side, due to selective and continuous removal of hydrogen from the reaction zone. In general, the presence of palladium particles in the membrane decreased cyclohexane conversion owing to the reverse reaction, and this effect was larger for PM2 than for PM1.

Yeung et al. (1996) studied ethane dehydrogenation to ethylene in an IMRCF using Pt-Sn/γ-Al$_2$O$_3$ catalysts. Three types of pellets were prepared with the same total metal loadings (1.1 wt% Pt, 1.3 wt% Sn): (a) uniform Pt distribution, (b) narrow surface-step Pt distribution (Pt layer thickness ~10% of pellet radius), and (c) wide surface-step Pt distribution (Pt layer thickness ~40% of pellet radius). In all cases, Sn (added to enhance ethylene selectivity) was distributed uniformly in the pellets. The membrane was a thin dense film of palladium deposited on a porous 316 stainless-steel tube that permits transport *only* of hydrogen. As shown in Figure 5.5, ethane conversion increases with residence time for both nonuniform distributions, and is two to three times higher than the equilibrium conversion. The highest conversion is attained by concentrating the Pt catalyst at the pellet external surface. This indicates that the influence of transport processes in the pellet is not negligible and their presence is detrimental to catalyst performance.

Szegner et al. (1997) studied experimentally and theoretically the same reaction on step-type Pt-Sn/γ-Al$_2$O$_3$ catalysts in a well-mixed isothermal IMRCF. Four types of pellets with different Pt distributions but with approximately the same amount of platinum and tin were used, as shown in Figure 5.6. The membrane used was a composite mesoporous γ-Al$_2$O$_3$ membrane. In order to minimize reactant loss, the effective permeability of the commercial composite Membralox™ membrane with 2-μm γ-Al$_2$O$_3$ membrane layer was reduced by slip-casting of alumina sol. The slip-cast modified membrane was approximately 17 μm thick (see Figure 5.7 for cross sections of original and slip-cast modified Membralox membranes). As shown in Figure 5.8, optimum reactor performance was obtained when the catalyst was concentrated at the surface of the pellet. In this case

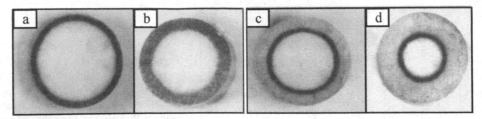

Figure 5.6. Pt-Sn/γ-Al$_2$O$_3$ pellets with different platinum distributions:

Part	Catalyst distribution	Active-layer center $\bar{s}$	Active-layer width 2Δ	Pt loading (wt%)	Sn loading (wt%)
a	Narrow surface step	0.95	0.10	1.11	1.36
b	Wide surface step	0.875	0.25	1.17	1.43
c	Near surface step	0.70	0.10	1.09	1.34
d	Deep subsurface step	0.50	0.10	1.16	1.35

(From Szegner et al., 1997.)

supraequilibrium conversions, about 80% beyond equilibrium values, were observed. Ethane conversion decreased as the catalyst location moved from the surface towards the pellet interior. This is due to strong diffusional resistance in the inert region before the reactant contacts the active sites within the catalyst pellets. Since dehydrogenation of ethane is a positive-order reaction, decreasing reaction rates are obtained as a consequence of the reactant concentration gradient. Hence, by placing the active layer as a narrow step at the pellet surface, intrapellet diffusion limitations are minimized, leading to higher reaction rates. It is worth noting that, owing to high intrinsic catalyst activity, conversions obtained using wide-surface-step pellets (Figure 5.6.b) were only slightly lower than with narrow-surface-step pellets (Figure 5.6.a). The performance of the IMRCF containing pellets with the deepest catalyst location (Figure 5.6.d) was inferior even to that of a fixed-bed reactor packed with surface-catalyst pellets. The theoretical model assumed isothermal, steady-state, well-mixed conditions on both sides of the membrane, and included diffusion–reaction in the pellets and Knudsen diffusion across the membrane. The agreement between numerical and experimental

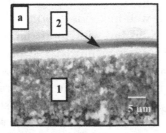 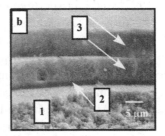

Figure 5.7. SEM micrographs of Membralox tubular membranes: (a) original and (b) slip-cast modified. (From Szegner et al., 1997.)

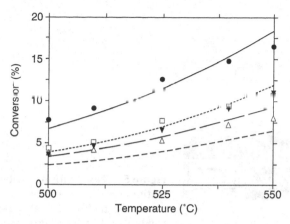

Figure 5.8. Effect of temperature on conversion of ethane in an IMRCF with platinum distributed as a narrow surface step (solid curve), near-surface step (long-dashed curve), and deep subsurface step (medium-dashed curve), and in an FBR with platinum distributed as a narrow surface step (short-dashed curve). The symbols and lines are the experimental data and model results, respectively. (From Szegner et al., 1997.)

results was good (see Figure 5.8), especially considering that no adjustable parameters were used, since intrinsic reaction kinetics and diffusivities were determined from separate experiments.

5.4.2 Preparation of Catalytic Membranes

For the preparation of catalytic membranes with nonuniform distribution of catalyst, conventional impregnation methods discussed in Chapter 7 cannot be applied, because the typical membrane thickness is too small (2–50 μm). In addition, it is difficult to prevent the penetration of catalyst precursor into the graded support material, which is undesirable. For example, in the impregnation of graded-structure alumina membranes (Membralox) with platinum precursors, it was observed that while most of the metal was deposited uniformly on the top γ-Al$_2$O$_3$ layer, rich in hydroxyl groups, part of it also penetrated the α-Al$_2$O$_3$ support (cf. Uzio et al., 1991).

Crack-free alumina membranes with nonuniform distribution of catalyst can be prepared by *sequential slip-casting* (Yeung et al., 1994, 1997). According to this technique, the membrane is built layer by layer, alternately slip-casting inert alumina and catalyst-containing sols to achieve the desired active layer width and location within the membrane. Some examples of catalytic membranes containing Pt are shown in Figure 5.9. Figure 5.9.a shows a 6-μm-thick γ-Al$_2$O$_3$ membrane, with a narrow 0.6-μm active layer, consisting of a 5-wt% Pt/alumina layer located 1.5 μm from the membrane surface. Figure 5.9.b is a micrograph of a 7.2-μm-thick

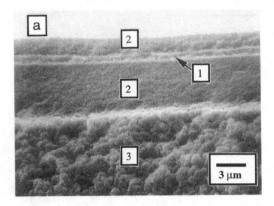

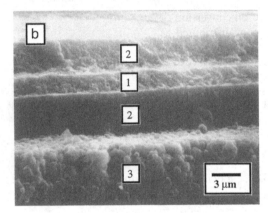

Figure 5.9. Nonuniform distribution of platinum catalyst in membranes prepared by the sequential slip-casting technique: 1, 5-wt% Pt/γ-alumina 2, γ-alumina 3, intermediate aluminas. (From Yeung et al., 1997.)

γ-Al$_2$O$_3$ membrane, with 1.5-μm-thick 5-wt% Pt/alumina located 3.3 μm from the membrane surface. The membrane top layer has a nominal pore size of 50 Å. Region 3 in both figures shows graded layers of a Membralox ceramic tube. These scanning electron micrographs show that precise control of the active-layer width and location can be attained using this technique. In addition, this preparation method restricts the catalyst to the active layer, preventing catalyst loss onto the graded ceramic support.

As mentioned in section 2.4.2, the relation of catalytic surface area vs loading is critical for determining the optimal distribution of catalyst, and this relation can be affected by the preparation conditions of the active sol. The active sol used for the catalytic membranes shown in Figure 5.9 is prepared by ion-exchanging chloroplatinic acid with a colloidal suspension of alumina (A1 sol: 25 wt% alumina, average particle size 150 nm; A2 sol: 20 wt% alumina, average particle size 50 nm). By maintaining the pH at 4 during ion exchange, a linear relationship between surface area and loading is obtained as shown in Figure 5.10. In this case, Pt dispersions are high, as shown in Table 5.1. If buffer is not used, and the pH is allowed to decrease, the relationship between surface area and loading is nonlinear, exhibiting an upper threshold of 0.75 m^2/g for Pt loading >1 wt%, and a consequent loss of dispersion. These results indicate that the presence of buffer leads to higher dispersion and a smaller Pt crystallite size.

Table 5.1. Platinum catalyst surface area, average crystallite size, and dispersion.

Pt loading (wt%)	Pt surface area (m²/g)	Pt crystallite size (nm)	Dispersion
Pt-A1 (pH = 4)			
0.1	0.13	2.12	0.53
0.5	0.58	2.43	0.47
1.0	0.51	5.50	0.21
1.7	0.97	4.92	0.23
2.5	1.69	4.13	0.27
5.0	5.02	4.64	0.24
Pt-A2 (no buffer)			
0.1	0.05	5.43	0.21
0.5	0.29	4.89	0.23
1.0	0.68	4.10	0.28
2.6	0.70	10.54	0.11
5.1	0.75	18.87	0.06

Different distributions of catalyst within the membrane can be obtained by varying the thickness of the slip-cast layers (both inert and active) as well as their arrangement. The widths of the layers are controlled by the slip-casting parameters, i.e. slip-casting time, viscosity, slip concentration, particle size, alumina concentration and support properties (i.e. pore structure, wettability). More specifically, the thicknesses of the inert (A1) and active (5-wt% Pt-A1) membrane layers vary linearly with the square root of slip-casting time as shown in Figure 5.11.a. This is consistent with the equation that Adcock and McDowall (1957) derived for the gel layer thickness L_g as a function of the system physicochemical parameters, following the Kozeny–Carman development for filter pressing:

$$L_g = \left(\frac{2 K_g \, \Delta P_g \, t}{\eta \, \alpha_g} \right)^{0.5}$$

(5.14)

where K_g is the permeability of the gel layer, ΔP_g is the pressure drop across the

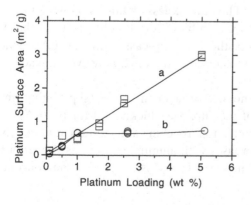

Figure 5.10. Platinum surface area as a function of platinum loading for a Pt-A2 sol: curve a, buffer pH = 4; curve b, no buffer. (From Yeung et al., 1997.)

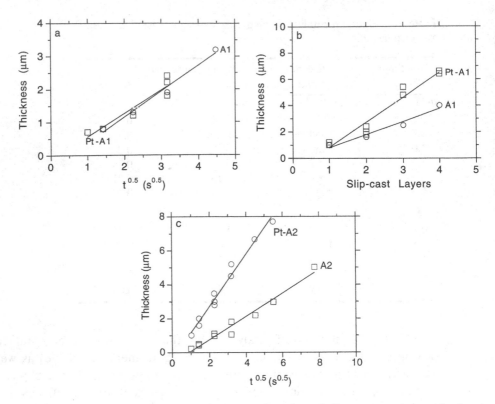

Figure 5.11. Membrane layer thickness as a function of slip-cast parameters for inert 2-wt% Al_2O_3 and active 5-wt% Pt/alumina –2-wt% Al_2O_3 sols. (a) Parameter: (time)$^{1/2}$; inert and active A1 sol. (b) Parameter: number of slip-cast layers; inert and active A1 sol, (c) parameter: (time)$^{1/2}$; inert and active A2 sol. (From Yeung et al., 1997.)

gel layer, t is time, η is the solvent viscosity, and α_g is a constant dependent on both the sol concentration and the support characteristics. Besides increasing the slip-casting time, the thickness of the membrane can also be increased by multiple slip-casting, as shown in Figure 5.11.b. In this experiment, it is observed that the inert and active sols display different behavior after the first slip-cast, because the support on which subsequent layers are deposited is different in the two cases. In Figure 5.11.c, the dependence of layer thickness on time is shown for inert (A2) and active (Pt-A2) sols. The A2 sols exhibit different behavior from the A1 sols (compare Figure 5.11.a and c). For the same slip-casting time, sol A1, which has larger alumina particles, gives a thicker layer than A2. However, the active layer Pt-A2 is thicker than Pt-A1.

Figure 5.12.a shows the dependence of layer thickness (slip-casting time 5 s) on the alumina concentration of the slip. The thicknesses for both A1 and A2 sols increase with alumina concentration, reaching a maximum at about 15 wt%. The viscosity of the slip also increases with alumina content as shown in Figure 5.12.b. Both sols display sharp increases in viscosity for alumina concentration

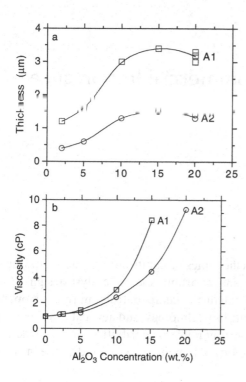

Figure 5.12. (a) Membrane layer thickness and (b) viscosity of alumina slips (A1 and A2 sols) as a function of alumina content. (From Yeung et al., 1997.)

above 10 wt%, which results in the thickness leveling off as observed in Figure 5.12.a. Sol viscosity is important in preventing pore clogging as well as controlling gelation rate of the slip-cast (Cot et al., 1988). The viscosity can also be controlled by addition of binders such as PVA, which help regulate the gelling rate and reduce crack formation in the final membrane.

6

Special Topics of Commercial Importance

I n this chapter, we consider specific topics of significant commercial value where nonuniform distribution plays an important role in catalyst design. These include catalysts for automotive exhaust cleanup, petroleum refining operations such as hydrotreating and cracking, biotechnology, and acid catalysis. Particularly in the case of automotive catalysis, nonuniform distribution of noble metals provides critical advantages for pollution abatement reactions and has been employed extensively.

6.1   Automotive Exhaust Catalysts

Automobile exhaust is considered the main source of air pollution in urban areas. The major pollutants in exhaust gas are carbon monoxide, hydrocarbons, and oxides of nitrogen. Following the Federal Clean Air Act of 1970, which called for a drastic reduction in these emissions, on all cars made in the United States since 1975, catalysts have been used to convert the pollutants into harmless gases. The catalysts in pre-1981 automobiles were *oxidation catalysts*, which controlled carbon monoxide and hydrocarbons only, by oxidizing them to form carbon dioxide and water. Their active components were platinum and palladium, which were deposited on substrates with a large surface area, either a monolith or pellets packed in shallow, pancake-shaped converters. Starting with the 1981 model year, because of stricter nitrogen oxide controls and fuel economy requirements, oxidation catalysts were replaced by *three-way catalysts* (TWCs), which simultaneously control all three of the major pollutants. They oxidize carbon monoxide and hydrocarbons while reducing nitrogen oxides as well. To perform these tasks, commercial TWCs contain platinum, palladium, rhodium, and cerium oxide as active components. Platinum and palladium provide activity for carbon monoxide and hydrocarbon oxidation, while rhodium is excellent for NO_x reduction. Since there are opposing requirements for oxidation and reduction catalytic reactions, the engine must operate in a narrow window around the stoichiometric air–fuel ratio where all three pollutants are removed efficiently (see Figure 6.1). Excellent reviews of the use

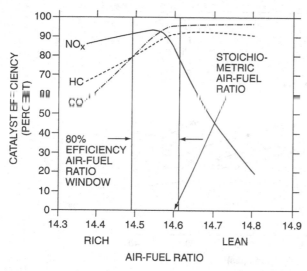

Figure 6.1. Steady-state conversion of carbon monoxide (CO), hydrocarbons (HC), and nitrogen oxides (NO$_x$) for a three-way catalyst as a function of the air–fuel ratio. (From Kummer, 1980.)

of catalysts in automotive exhausts are available in the literature (Kummer, 1980; Taylor, 1984; Heck and Farrauto, 1995).

6.1.1 Design of Layered Catalysts

Under normal operating conditions the catalyst is at sufficiently high temperature so that reactions are diffusion-controlled, and for this reason eggshell distribution is beneficial. However, a complication arises from the occurrence of *deactivation* processes: poisoning due to the presence of trace quantities of phosphorus, sulfur, and metals (Pb, Zn, etc.) in the engine oil and the fuel, and thermal sintering. Phosphorus and metal poisoning are essentially irreversible, while poisoning by sulfur compounds is reversible (Summers and Hegedus, 1979). In addition, the catalyst has to be resistant to frequent thermal cycles as well as mechanical vibrations. Since the size of the automotive catalyst market is large (see section 1.1) and the noble metals are expensive, optimization of their use is of particular interest.

Some of the earliest applications of subsurface step-type catalysts were for automotive exhaust emission control. Michalko (1966a,b) and Hoekstra (1968) patented noble-metal catalysts with subsurface location of the active components, which showed improved long-term stability by delaying catalyst poisoning. The poisoning process is nonselective, i.e., it involves both the support and the active catalyst. Furthermore, for phosphorus and metal poisoning it is diffusion-controlled, i.e., the poison penetrates the catalyst support in the form of a sharp front, whose velocity is independent of the local concentration of the active elements (Hegedus and Summers, 1977). Under these conditions, the support can

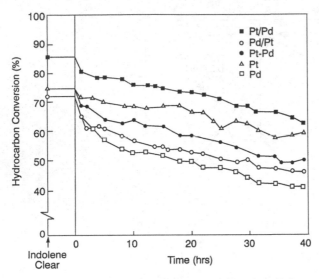

Figure 6.2. Hydrocarbon conversion as a function of time, during accelerated poisoning experiments with pellets having different catalyst distributions. Pt/Pd: Pt(exterior)/Pd(interior); Pd/Pt: Pd(exterior)/Pt(interior); Pt–Pd: Pt and Pd (both exterior); Pt: Pt(exterior); Pd: Pd(exterior). (From Summers and Hegedus, 1978.)

be used as a poison-getter to protect the unpoisoned noble metals beyond the poisoned zone.

Summers and Hegedus (1978) studied the case where only the two oxidizing reactions occur, and compared several distributions of Pt and Pd: Pt(exterior)/Pd (interior), Pd(exterior)/Pt(interior), Pt–Pd(exterior), Pt(exterior), Pd(exterior). Subsurface impregnation of both Pt and Pd would offer the best protection from poisoning, but was not considered due to increased mass transfer resistances. Hydrocarbon conversions during accelerated poisoning experiments for the various distributions are shown in Figure 6.2. The catalyst pellet which was impregnated with Pt at the outer shell and with Pd in a separate subsurface band (Pt/Pd) exhibited the best performance before, as well as after, poisoning. This was attributed to the different poison characteristics of Pt and Pd. Pt is more resistant to poisoning, with a nonzero residual activity, while Pd completely loses its activity once the poisoned shell penetrates through the Pd-impregnated shell. The Pt–Pd coimpregnated catalyst (Pt–Pd) showed unsatisfactory performance because the two elements form an alloy with Pd on its outer surface, so that coimpregnated Pt–Pd catalysts tend to poison with the characteristics of Pd. For CO oxidation, the fresh Pt–Pd catalyst showed the highest activity, but after accelerated sintering or poisoning experiments the Pt/Pd catalyst exhibited the best performance, because Pd was protected from poisoning by subsurface impregnation.

The above *multilayer design* was extended to three-way catalysts by Hegedus et al. (1979a) (see also Hegedus and Gumbleton, 1980). These catalysts included Rh for NO_x reduction, and its high cost compared to that of Pt and Pd, as well as

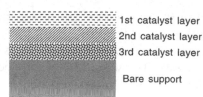

1st catalyst layer

2nd catalyst layer

3rd catalyst layer

Figure 6.3. Multilayer catalyst design concept.

Bare support

its strong sensitivity to phosphorus and lead poisoning, suggests using it in small quantities with protection from poisoning. Based on these considerations, in each pellet the noble metals were located in three separate layers, one underneath the other, as illustrated in Figure 6.3. The first, at the pellet's external surface, was the Pt layer whose depth was determined so as to entirely contain the expected penetration depth of the poison during the desired catalyst lifetime. Next was located the Rh layer, so that diffusional resistances for NO_x reduction would be minimized and Rh would also be protected from poisoning. Finally, the Pd layer was placed at the deepest location, providing good lightoff performance, especially for aged catalysts, because of its good thermal stability. This configuration also prevented undesirable alloy formation among the noble metals. While maintaining the same noble metals and their order in the layer configuration, Vayenas and coworkers (Vayenas et al., 1994; Papadakis et al., 1996) have proposed the use of different supports that further enhance catalytic activity. Multilayer three-way catalyst design remains an active area of research, and numerous publications continue to appear in the patent and scientific literature.

There is a strong motivation to avoid using Rh due to its high cost, and hence research has been directed towards alternative catalyst formulations. However, Pd- and Pt-based catalysts are typically not as effective for NO conversion. Some researchers have tried to overcome these disadvantages by using various additives. Hu et al. (1996) improved the performance of Pd-only three-way catalysts using a two-layer design, where the first layer contained Pd and the second Pd and ceria. Both layers utilized alumina as the support and contained base-metal oxide stabilizers. The Pd layer provided high activity at low temperatures, while the Pd–Ce layer offered large oxygen storage capacity, which gave improved three-way activity at high temperatures. The two-layer Pd catalyst exhibited performance comparable to a Pt/Rh catalyst.

Automotive exhaust catalysts constitute a remarkable example of catalyst design despite the fact that in many instances, because of complexity of the problem and lack of detailed expressions for the kinetics of the various reactions involved, the design was based on experimental observations rather than on quantitative models.

6.1.2 Nonuniform Axial Catalyst Distribution

It has been experimentally demonstrated by various investigators (cf. Komiyama and Muraki, 1990; Beck et al., 1997) that poisons tend to deposit preferentially near the monolith inlet. Hence, two-zone monolith configurations have been suggested, where the upstream portion of the monolith is left unimpregnated. Oh and Cavendish (1983) investigated theoretically the poison resistance of such monoliths and found that the warm-up time of a poisoned monolith was lowest for an intermediate length of the inert first zone. This was due to balance between insufficient protection of the noble metal at short inert-zone lengths and long time to reach reaction temperature at long lengths.

Two-zone monoliths have been shown to offer improvements in other respects as well. Wu and Hammerle (1983) studied two-zone converters where the first zone contained only Pd, while the second zone contained Pt and Rh. This combination showed improved thermal resistance and lightoff performance while offering a reduction in precious-metal cost. The reason was that activity of the second zone could be protected from thermal damage by the presence of the first zone, as long as the latter had sufficient thermal resistance to maintain some activity after repeated high-temperature exposure. Accordingly, Pd was placed in the first zone because it is more resistant to sintering under oxidizing conditions than Pt and Rh.

Another area where nonuniform catalyst distribution can be beneficial is reduction of *cold-start emissions*. In the first 1–2 min after cold start of an automobile, the catalyst temperature is too low for the reactions to take place. As a result, significant amounts of pollutants pass unconverted through the catalyst. They constitute a large fraction of the total tailpipe emissions. Nonuniform catalyst distribution along the monolith length can result in shorter warm-up period and thus lower cold-start emissions. Oh and Cavendish (1982) examined the lightoff behavior of three axial catalyst distributions: uniform, linear increasing, and linear decreasing. They found that CO cold-start emissions are reduced when the noble metal is concentrated in the upstream section of the monolith, as shown in Figure 6.4. If strong poisoning is present, this beneficial effect is counterbalanced by the preferential poison accumulation in this region, so that it may be preferable to concentrate the noble metal in the downstream section of the monolith where poison

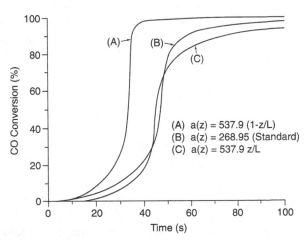

Figure 6.4. Effect of axial catalyst activity profile on lightoff. (From Oh and Cavendish, 1982.)

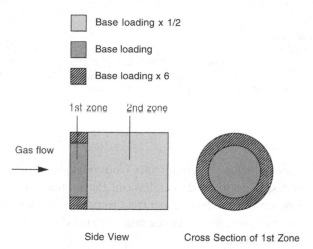

☐ Base loading x 1/2

■ Base loading

▨ Base loading x 6

Figure 6.5. Two-dimensional catalyst distribution concept. (From Naoki et al., 1996.)

concentration is low (Oh and Cavendish, 1983). Psyllos and Philippopoulos (1993) numerically studied the performance of monoliths with various parabolic axial catalyst distributions, and showed that certain distributions have shorter warm-up periods than the uniform one. Tronci et al. (1999) demonstrated theoretically that further improvements can be obtained by optimizing the axial catalyst distribution. By using the optimal distribution, lightoff time was decreased significantly as compared to the uniform distribution. A two-zone converter was also investigated, and resulted in similar performance to that of the optimal distribution.

A two-dimensional catalyst distribution pattern, both axially along and radially across the monolith, was investigated by Naoki et al. (1996), and is shown in Figure 6.5. The quantity of noble metal was six times the base loading in the outer region of the first zone, equal to the base loading in the inner region of the first zone, and one-half the base loading in the second zone. This catalyst distribution was shown theoretically to exhibit higher conversion efficiency during cold start than did a uniform monolith with the same total catalyst amount. The investigators considered radial variations in inlet velocity, inlet temperature, radial heat transfer, and diabatic operation. High loading in the first zone was found to improve cold-start behavior, in agreement with the discussion above. Concentrating the catalyst towards the central region of the first zone improved conversion at the early stage of warm-up, while concentrating it towards the periphery, as shown in Figure 6.5, increased conversion at later stages of the warm-up, after lightoff of the central region. The latter design exhibited the best performance and could be improved further by insulating the monolith.

6.2 Hydrotreating Catalysts

Hydrotreating catalysts are used in the petroleum industry to remove various compounds (e.g. metals, sulfur) from the bottom fraction of distilled crudes. This

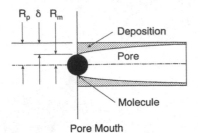

Figure 6.6. Schematic representation of pore-mouth plugging of catalyst during hydrodemetallation. (From Limbach and Wei, 1988.)

is required in order to prevent these contaminants from poisoning the catalysts used in subsequent processes. In hydrodemetallation, the metals are removed by irreversible adsorption on the catalyst. In contrast to previous situations, this deactivating process is actually desired, since the goal is to maximize the amount of metals deposited on the catalyst.

Due to strong diffusional limitations, metal deposition primarily occurs at the pore mouth, which hinders the entrance of other reactant molecules, thus leaving a substantial fraction of the catalyst in the internal part of the pellet unutilized. This process is illustrated schematically in Figure 6.6, where it is shown that when the thickness of the deposit equals the difference between the pore radius R_p and the reactant molecule radius R_m, then the transport through the pore mouth is inhibited and the pellet becomes inactive. For first-order deposition kinetics, the time t when such pore-blocking occurs can be estimated from the relationship

$$t = \rho_d(R_p - R_m)/kC \tag{6.1}$$

where ρ_d is the deposited layer density, C is the reactant concentration (kg/kg_{oil}), and k is the rate constant of the deposition process based on catalyst surface ($kg_{oil}/m^2 \cdot s$).

In order to use the catalyst more effectively, various strategies have been explored, involving optimization of pore size distribution and of catalyst shape. In a detailed theoretical study, Limbach and Wei (1988) investigated the performance of catalyst pellets with nonuniform activity distributions, using a diffusion–reaction model. Hydrodemetallation was simulated by a single isothermal first-order reaction. It was assumed that the intrinsic reaction rate per unit active surface area is not affected by deposition. The change of active surface area as well as the decrease of the effective diffusion coefficient with deposition was accounted for, because, as illustrated in Figure 6.6, the deposited metal decreases the cross-section area of the pores.

In the uniform catalyst pellet, the pinch point (where pore plugging occurs) is located at the pore mouth. In this case, the interior portion of the catalyst is not properly utilized. Therefore, in order to maximize the total metal deposit, the pellet must fill from the center outwards. This can be accomplished using an initial catalyst distribution increasing towards the pellet center, which shifts the pinch point from the pellet surface towards the center. This catalyst distribution also compensates for low reactant concentration in the pellet interior (the Thiele modulus of hydrodemetallation is typically large), and increases the amount of

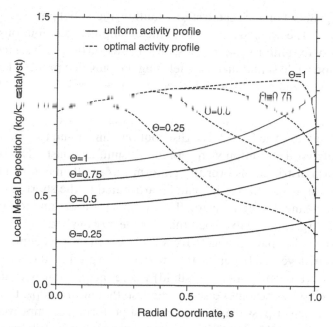

Figure 6.7. Metal deposition profiles resulting from optimal and uniform activity profiles at various values of dimensionless operating time Θ. Dimensionless time is based on useful catalyst lifetime. (From Limbach and Wei, 1988.)

deposition. No restrictions were placed on either the local or the overall catalyst loadings in seeking an optimal solution that maximizes the amount of contaminant metal deposition over the useful catalyst life. The optimal catalyst distribution was calculated numerically. The performance of a nonuniform catalyst was compared with that of a uniform catalyst which was already optimized for catalyst loading and pore diameter. As expected, the obtained optimal distribution exhibited a sharp maximum towards the center of the pellet followed by a decrease towards the external surface. This distribution provided a substantial increase in the overall amount of metals deposited, as shown in Figure 6.7. As approximations of the optimal distribution, edge-reduced (where the catalyst distribution was uniform, but reduced only at the outer edge of the pellet) and egg-yolk catalysts were also considered. It was found that the total metal deposited in a uniform pellet could be increased by nearly 25% utilizing the edge-reduced profile.

Chiang and Tiou (1992) extended the above ideas to the fixed-bed reactor. In order to explore the effects of nonuniform catalyst distribution, they considered three specific cases: linear (increasing towards the center), edge-reduced, and egg-yolk catalysts, with the same total loading. The evaluation was based on the reactor metal deposition capacity at the point where the metal compound concentration in the reactor effluent reached 30% of the inlet value. The linear profile exhibited the largest improvement in metal deposit (almost 50%) over the uniform distribution. The edge-reduced and egg-yolk catalysts also showed significant improvements over the uniform pellets. For each catalyst distribution, the initial pore size of the pellet was optimized. The optimal pore sizes of the nonuniform distributions were smaller than those of the uniform one, because they increased the pore surface area for metal deposition while simultaneously delaying the plugging process. Further improvement was realized with two- and three-zone reactors, but the additional improvement was smaller for the nonuniform distributions.

The case of a moving-bed reactor was also investigated (Chiang and Fang, 1994), using two families of active catalyst distributions: linear and two-step. Pore size and active catalyst distribution were optimized simultaneously. The benefit of using nonuniform distributions over the uniform was small, in contrast to the fixed bed reactors discussed above, and can be attributed to the short residence time of catalyst in the moving-bed reactor. The optimal pore size and total metal deposit were both larger for the fixed bed than for the moving bed.

In all the above studies pore connectivity was not taken into account. Arbabi and Sahimi (1991a,b) developed a percolation model which predicts the occurrence of a transition point (the percolation threshold) where the previously connected cluster of unplugged pores becomes disconnected, so that macroscopic transport and reaction are no longer possible. Similar catalyst performance improvements obtained by Limbach and Wei (1988) for nonuniform catalyst distribution were also observed. Nonuniform catalyst distributions with activity increasing towards the pellet center lead to increased metal deposition capacities.

For hydrodesulfurization (HDS), Asua and Delmon (1987) investigated theoretically the behavior of catalysts containing molybdenum and cobalt sulfides. In this case, catalyst performance optimization can be pursued by varying both the overall catalyst concentration (i.e. Mo+Co) as well as its composition (i.e. Mo/Co). They used a detailed kinetic model which accounted explicitly for the catalyst composition, as well as for the concentration of active sites, and showed that the catalyst performance could indeed be improved by using nonuniformly distributed catalyst pellets.

Goula et al. (1992a) studied experimentally the hydrodesulfurization of thiophene on eggshell, egg-white, egg-yolk, and uniform MoO_3/Al_2O_3 catalysts. All catalysts had similar loadings (12–14% MoO_3) and dispersions (0.098–0.134), but different numbers of active sites. The highest activity for HDS was found for the egg-white catalysts, and it was related to the number and type of active sites rather than the Mo distribution. However, from powder experiments it was established that diffusional resistances were present, and the effectiveness factor followed the trend $\eta_{eggshell} > \eta_{egg\text{-}white} > \eta_{egg\text{-}yolk}$. On the other hand, the selectivity towards butane, which is formed by hydrogenation of the unsaturated hydrocarbons produced by HDS, was affected by the catalyst distribution and followed the trend $S_{egg\text{-}yolk} > S_{egg\text{-}white} > S_{eggshell}$. This behavior was related to the residence time of unsaturated hydrocarbons inside the pellets. In addition, the uniform catalyst exhibited the lowest activities for both hydrodesulfurization and hydrogenation, and this was attributed to partial clogging of the pores during preparation.

6.3 Composite Zeolite Catalysts

Composite zeolite catalysts consist of zeolite catalyst particles embedded in a support (e.g. silica–alumina), which may exhibit some modest catalytic activity itself. The support has much larger pores, and therefore diffusion is faster in the support than in the active catalyst. This is the main advantage of composite catalysts, since the reactants have easier access to the active catalyst particles than in the case where the entire pellet is composed of the active catalyst alone (Ruckenstein, 1970; Smirniotis and Ruckenstein, 1993). In the presence of pore-mouth poisoning, these systems exhibit a further advantage, which has been discussed by Varghese and Wolf (1980) with reference to uniformly distributed pellets. It was shown that by increasing the poison diffusivities and poison-getter capacity of the support, the catalyst lifetime can be increased relative to pellets which contain the active catalyst alone.

Dadyburjor (1982) investigated the possibility of improving composite pellet activity through the use of nonuniform catalyst distribution. In contrast to the above studies, the matrix was considered to exhibit some catalytic activity. However, it was lower than that of the embedded particles, as is the case for hydrocarbon catalytic cracking on zeolite particles embedded in silica–alumina. The diffusivity in the matrix was larger than in the catalyst particles, and a first-order reaction was considered. The simulations indicated that the overall reaction rate was in the order uniform > concentrated egg-yolk > concentrated eggshell, where "concentrated" indicates that the catalyst layer is undiluted zeolite. This behavior is a consequence of the matrix reactivity. The matrix in fact contributes significantly to the overall reaction rate for the uniform and egg-yolk but not for the eggshell catalyst, because in the last significant diffusional resistances are introduced by the external zeolite layer.

The effect of a uniform deactivation process, mainly due to coking, was also investigated using the "snapshot" approach, i.e. identifying the optimal zeolite distribution at various deactivation levels, which were simulated by using progressively smaller zeolite reactivity and matrix diffusivity. For a fresh catalyst, the optimal distribution was monotonically decreasing from the pellet center to zero at the pellet surface. When diffusional resistances were increased, either by increasing the pellet size or by simulating the effect of coking, the optimal distribution changed shape and exhibited a maximum at a subsurface location. For a first-order isothermal reaction, the eggshell catalyst would give maximum conversion if the zeolite and matrix diffusivities were the same. However, for this system there is an interplay between two levels of diffusion resistances and reactivities. At smaller pellet sizes, the higher reactivity of zeolites overshadows its lower diffusion coefficient. At larger pellet sizes, it is more important for the matrix to allow the reactant to diffuse inside the pellet than for the zeolite to convert as much as possible of the reactant at the pellet surface. However, at least for the explored parameter values, the optimal distribution resulted only in modest improvement *in overall reaction rate* over the uniform pellet.

The improvement *in selectivity* through the use of nonuniform active catalyst distribution was larger, and ranged from 10% to 100% relative to the uniform distribution for the range of parameters investigated (Dadyburjor, 1985). Parallel and series reaction systems with first-order kinetics were examined to represent hydrogen transfer, β-scission, and coke formation, which typically occur during catalytic cracking. The performance of concentrated eggshell, egg-white, and egg-yolk and of uniform and diluted egg-white distributions was examined. In the last case, the internal layer was formed by a mixture of zeolite and matrix (silica–alumina). In all distributions, the total amount of zeolite was constant. Optimization was carried out within the family of diluted egg-white catalysts by computing the values of inner and outer radii of the internal layer leading to maximum selectivity or yield. Note that by properly selecting the two radius values, the diluted egg-white distribution reduces to each of the other four distributions mentioned above. In the *parallel reaction* system where the desired product is formed by hydrogen transfer and the undesired product by β-scission, the concentrated eggshell distribution offered the highest selectivity, while diluted eggshell distributions optimized yield. In the *series reaction* scheme, A→C→D, where the desired product C is formed by hydrogen transfer and further reaction results in undesired coke precursors D, selectivity was maximized by diluted eggshell distributions. For cases where a diluted eggshell distribution was optimal, the thickness of the diluted layer became smaller with deactivation.

Lee and Ruckenstein (1986) represented the catalytic cracking process using the reaction scheme

$$A \rightarrow B \rightarrow C \atop \searrow \swarrow \atop D \qquad (6.2)$$

where component A is oil, C is gasoline (which is the desired product), D is gases which result from the secondary cracking of gasoline, and B is an intermediate compound. If formation of the intermediate B is neglected, the above network simplifies to a series reaction scheme. In this case the rate constant for A→C was taken larger in the zeolite than in silica–alumina, while the rate constant for C→D was taken smaller in the zeolite than in silica–alumina. It was found that a diluted eggshell distribution of the zeolite was preferable to a uniform or egg-yolk distribution for yield and selectivity maximization. This leads to a high rate of formation of C at the zeolite location, and a significant part of C can diffuse out of the pellet without being transformed to D in the core region. As mentioned above, a similar diluted eggshell distribution was found to be optimum for selectivity maximization in a consecutive reaction scheme by Dadyburjor (1985), even though both rate constants were larger in the zeolite than in the matrix. This indicates that if the rate constant for the desired product is greater in the zeolite than in the matrix, the optimal distribution type is the same regardless of the rate constant of the secondary reaction being larger or smaller in the zeolite.

When the intermediate B is included in the reaction set, then for the case where the rate constants of all reactions but B→C are greater in the silica–alumina matrix than in the zeolite, the diluted egg-yolk distribution leads to higher yield and selectivity than the uniform distribution. This occurs because the formation of component B in the outer region can lead to a greater amount of desired C in the inner region. The optimum radius of the diluted core and the optimum yield both depend on the values of the parameters. When the overall volume fraction of silica–alumina was also considered as an optimization variable, it was found that yield exhibited a maximum with respect to both the diluted-core radius and the volume fraction of silica–alumina.

Dadyburjor and coworkers (Martin et al., 1987; White and Dadyburjor, 1989; Dadyburjor and White, 1990) studied extensions of the above schemes including changes in reaction orders as well as position-dependent rate constants and diffusivities arising from nonuniform poisoning. Since these variations induce complex interactions between transport and reaction in the zeolite and matrix, a variety of zeolite distributions were found to be optimal, depending on the specific case investigated.

6.4 Immobilized Biocatalysts

After immobilized enzymes were successfully industrialized by the Tanabe Seiyaku Company in 1969 (Tosa et al., 1969), utilization of biocatalysts such as enzymes, microorganisms, and cells increased sharply. Many papers have since been published regarding immobilization methods and applications of immobilized biocatalysts, and various processes have been developed at an industrial scale (cf. Furusaki, 1988; Chibata et al., 1992; Lilly 1996). Mass transfer resistance in immobilized enzymes can be significant because the reactions take place in liquid (usually aqueous) phase and reactant transport inside the support is generally slower than reaction. The importance of such resistance for catalyst and reactor performance has been widely recognized in the biochemical engineering literature (cf. Vieth et al., 1973; Radovich, 1985; Buchholz and Klein, 1987; Furusaki, 1988).

The most common kinetic expression to describe a biochemical reaction is the Michaelis–Menten kinetics,

$$r = \frac{r_{max}C}{K_m + C} \tag{6.3}$$

where r_{max} is the maximum reaction rate and K_m the Michaelis–Menten constant. This kinetics is monotone increasing in substrate concentration, behaving as first-order at low and as zero-order at high concentrations. Thus, for isothermal conditions, mass transfer resistance can only be detrimental. Based on the theoretical results discussed in Chapter 2, we expect the optimal distribution to be a surface Dirac-delta, and its advantage over a uniform distribution is more significant at low substrate concentration and large Thiele modulus. Accordingly, most studies which consider nonuniform catalysts have focused on eggshell distributions.

One of the early studies was by Horvath and Engasser (1973), who investigated theoretically the properties of pellicular heterogeneous catalysts, which consist of a fluid-impervious core supporting a spherical annulus of the catalytically active porous medium. This configuration is equivalent to an eggshell enzyme distribution. For Michaelis–Menten kinetics under isothermal conditions, it was shown that a fixed amount of enzyme gives higher overall reaction rate when it is confined to the outer shell than when it is distributed uniformly in a porous pellet of the same diameter. As expected, significant improvements in overall reaction rate are realized by the eggshell catalyst for high Thiele modulus and sufficiently low reactant concentration. The latter occurs at the end of batch operation, when high reactant conversions are required. These conclusions were confirmed experimentally for eggshell and uniformly distributed agarose–micrococcal nuclease in the hydrolysis of a phosphate compound, which follows Michaelis–Menten kinetics (Guisan et al., 1987). Under conditions of low reactant concentration and high Thiele modulus, the highest increase of effectiveness factor obtained theoretically was 35%, while a 16% improvement was observed experimentally.

Higher effectiveness factors for eggshell immobilized trypsin on derivatized glass carriers, as compared to a uniform distribution, were demonstrated experimentally and theoretically by Buchholz (1979) and Borchert and Buchholz (1979, 1984). For the hydrolysis of low concentration N-α-benzoyl-L-arginine ethyl ester solutions, experimental effectiveness factors for eggshell distributions were higher by up to 50%; when the higher-molecular-weight reactant casein was used, the effectiveness factor was higher by up to 100%, and this was attributed to large intraparticle concentration gradients. The improvements in effectiveness were more significant when larger support particles were used, since the Thiele modulus was then higher.

Because of the sensitivity of enzyme activity to pH, another reason for inferior performance of the uniform distribution could be slow proton diffusion from inner core of the particle to solution. Carleysmith et al. (1980a) showed that for deacylation of benzylpenicillin catalyzed by penicillin acylase, slow proton diffusion resulted in low reaction rates even when reactant diffusional limitations were decreased by the use of small particles. It was postulated that a pH gradient existed in the beads even in the presence of a buffer. On the other hand, for eggshell distributions, all the enzyme probably operated near its optimum pH.

Subsurface enzyme distributions can confer a measure of protection against degradation by local concentrations of harsh reagents which may need to be added during reaction. Carleysmith et al. (1980b) showed experimentally for the hydrolysis of benzylpenicillin that when the enzyme (penicillin acylase) was immobilized in a subsurface layer, the particles retained their initial activity to a larger extent after three consecutive batch reactions than with a shell-type enzyme distribution. This was attributed to the fact that the enzyme was protected from localized regions of high pH that resulted when alkali was added to maintain the pH of the reaction medium constant.

Park et al. (1981) extended Horvath and Engasser's work by studying theoretically the behavior of eggshell, egg-yolk, and uniform enzyme distributions for

Michaelis–Menten kinetics with substrate inhibition

$$r = \frac{r_{max}C}{K_m + C + K_{is}C^2} \tag{6.4}$$

and product inhibition

$$r = \frac{r_{max}C}{K_m[1 + (C_{product}/K_{ip})] + C}. \tag{6.5}$$

For product-inhibited reactions the eggshell distribution gave a higher effectiveness factor, similarly to the kinetics (6.3) – as is expected, since both rates show the same monotonic dependence on reactant concentration. In addition, the kinetics (6.5) exhibits a negative-order dependence on product concentration. This also favors eggshell catalysts, because the product can diffuse easily to the bulk fluid and hence its concentration in the enzyme layer is kept small. The substrate-inhibited kinetics (6.4) exhibits a reaction order with respect to substrate which varies from +1 at low to −1 at large concentrations. Thus, for certain conditions, the best catalyst performance is given by egg-white or egg-yolk distributions, since the diffusion barrier of the outer inert section reduces the inhibition effect. Reactions with similar kinetics were studied theoretically by Juang and Weng (1984) in pellets with increasing, decreasing (towards the center), and uniform enzyme distributions. As expected, the decreasing profile gave the highest effectiveness factor for the kinetics (6.3), while the increasing one was more effective for the substrate-inhibited kinetics in certain ranges of Thiele-modulus values. Following similar arguments to those in section 2.1.1, it was shown (Morbidelli et al., 1984b) that the *optimal* enzyme distribution is given by a Dirac-delta. For the kinetics (6.3) it is always located at the particle external surface, while for the substrate-inhibited kinetics (6.4) it may be located at any position within the support, depending on the reaction conditions. This finding is in agreement with the results discussed above for different types of nonuniform distributions.

Cases where *enzyme deactivation* was present were investigated for the kinetics (6.3) by Juang and Weng (1984) and Hossain and Do (1987). The decay of the enzyme activity also depended on substrate concentration, following a similar kinetic expression. When operating in the diffusion-limited regime, distributions with decreasing (towards the center) activity showed the best performance at short operation times, but for longer periods, higher effectiveness factors were obtained with uniform or increasing enzyme distributions. This occurs because as time elapses, the immobilized enzyme near the surface deactivates due to high substrate concentration, thus allowing more substrate to diffuse into the interior of the particle. Consequently, enzyme in the interior of the particle, which was previously unutilized because of lack of substrate, gradually plays a role in the reaction process. According to the analysis of selective deactivation systems (section 4.2), the optimal enzyme distribution should be a Dirac-delta. For short operating times and the kinetics (6.3), a surface Dirac-delta should provide the best overall performance, and this is consistent with the above findings (Juang and Weng, 1984; Hossain and Do, 1987). The fact that uniform and increasing distributions show higher conversions than decreasing distributions at large operating times indicates that the optimal enzyme concentration is an appropriate subsurface Dirac distribution.

Chung and Chang (1986) compared theoretically the performance of nonuniform biocatalysts for first-order kinetics in the presence of two different deactivation mechanisms representing the effects of temperature and time. The first was based on a reversible thermal denaturation model, according to which enzyme activity was an exponentially decreasing function of temperature (Ollis, 1972). This, combined with the usual Arrhenius dependence, gives rise to a maximum in the first-order reaction rate constant as a function of temperature. For the second deactivation mechanism, the reaction rate constant was assumed to decay exponentially with time. Three distributions with the same amount of enzyme were considered: uniform, eggshell, and egg-yolk. Egg-yolk immobilized enzymes had the smallest effectiveness factors, but they showed the best stability under temperature fluctuations. The egg-yolk distribution also exhibited better stability than eggshell catalysts over time, because in the former the effect of rate-constant decrease is counterbalanced by better utilization of the enzyme arising from the concomitant decrease of the Thiele modulus. Thus, stability with time increased in the order eggshell to uniform to egg-yolk.

6.5   Functionalized Polymer Resins

Improvement of catalyst performance through an appropriate nonuniform distribution of active element has also been investigated in the case of cation-exchanged polymer resins. They were originally developed for ion-exchange processes, such as those involved in water treatment or hydrometallurgy. These resins consist of a crosslinked polymer matrix, which is functionalized by acid groups covalently bonded to the polymer. The most common are made of a styrene–divinylbenzene copolymer, whose degree of crosslinking increases with divinylbenzene amount (usually 2–20 wt%). Functionalization is achieved by attaching sulfonic groups to the pendant phenyl groups. Due to the acid characteristics of the former, these ion-exchange resins can be used as supported acid catalysts for a variety of chemical reactions, including esterification, etherification, dehydration, alkylation, and isomerization (cf. Sherrington and Hodge, 1988).

As compared to the common homogeneous catalytic processes involving mineral acids such as sulfuric and hydrochloric acids, functionalized resins offer the advantages of heterogeneous catalysts: elimination of waste disposal problems, and catalyst reuse. However, it is worth noting that the activities and selectivities of the two catalytic systems are in principle different and depend on the specific reacting system and operating conditions.

6.5.1   Preparation of Nonuniformly Functionalized Resin Particles

Klein et al. (1984) showed that it is possible to prepare polymer particles with nonuniform distribution of sulfonic groups. The form of distribution depends on resin type (gelular or macroreticular) and sulfonation conditions. *Gelular* resins, which are in the form of compact beads of crosslinked polymer with no permanent porosity, can be penetrated by low-molecular-weight species only by swelling. This is a selective process, where diffusion and adsorption of species with higher affinity for the resin are favored. Thus, nonsulfonated resins preferentially sorb organic species, while sulfonated resins preferentially sorb polar species. This is important in organic synthesis, since it leads to different compositions of the reacting mixture in the external fluid phase and inside the resin particle, i.e. in the reaction locus. Therefore, by controlling the amount and distribution of sulfonic groups in the resin, one can control the selective sorption process and hence the composition in the reaction locus, which ultimately affects reaction rate and selectivity. The above clearly indicates that the performance of functionalized resins may differ significantly from that of homogeneous catalysts.

Macroreticular resins consist of a large number of gelular microparticles (~100 nm in diameter) separated by macropores through which diffusion takes place. Each of the gelular microparticles can swell, leading to selective sorption as discussed previously. On the other hand, the diffusion through the macropores is not selective, since intraparticle and external fluid phases are similar in composition. In the presence of chemical reaction, competition between diffusion and reaction arises, which can lead to concentration gradients of the reacting species in the resin particle, i.e. in the macropores. In addition, localized concentration gradients inside the microparticles develop due to the selective sorption process.

The swelling and diffusion–reaction phenomena described above take place during the sulfonation process, which is utilized to prepare different types of nonuniformly functionalized resin particles. For a gelular particle, surface distribution of acid sites is obtained by sulfonating the resin without a swelling agent. When such an agent is used, the intraparticle profile of sulfonic groups is the result of the relative rates of the sulfonation reaction and of diffusion of the sulfonating agent driven by the swelling. If the diffusion is slower, decreasing concentration profiles of the sulfonic groups from surface to center are obtained. In the opposite case, as well as for sufficiently long sulfonation times, a uniform distribution results.

The situation is more complex for macroreticular resins. As discussed above, different concentration profiles can develop inside the macroparticle (i.e. in the macroporous phase, and therefore also on the microparticle surface at the same location) and the gelular microparticles. In particular, Klein et al. (1984) obtained four different pairs of profiles in the macro- and microparticles, as shown schematically in Table 6.1. In the first case (type A), which corresponds to typical commercial resins, the distribution was uniform at both scales. This was obtained by using a suitable swelling agent and sufficiently long sulfonation times. In the remaining three cases, swelling agents of increasing strength were used to increase the characteristic diffusion time in the microparticles up to the point where it became comparable to that in the macropores. Thus, for type B particles, where H_2SO_4 as sulfonating agent and CH_3NO_2 as swelling agent were used, diffusion in the microparticles

Table 6.1. Nonuniform distribution of sulfonic groups in resin macroparticles and microparticles (From Klein et al., 1984.)

Distribution		Synthesis	Type	Figure
Macroparticle	Microparticle			
		H_2SO_4 completely sulfonated	A	
		H_2SO_4/CH_3NO_2	B	6.8
		$H_2SO_4/C_2H_4Cl_2$	C	6.9.a
		SO_3/N_2	D	6.9.b

was much slower than in the macropores. Therefore a uniform profile was obtained in the macroparticle, while the microparticles exhibited a surface distribution. For particles of type C, using $C_2H_4Cl_2$ as swelling agent, decreasing profile towards the center was obtained at both scales. Finally, for particles of type D, where the most active sulfonating agent SO_3 was used (in a gas-phase fluidized bed with nitrogen) together with a relatively short sulfonation time, the sulfonic group concentration profile was steeper in the macropores than in the microparticles.

Verification of the distributions discussed above was obtained by measuring the concentration of sulfonic groups by energy-dispersive X-ray analysis (EDX) of cross-sectional areas of resin particles (Bothe, 1982; Klein et al., 1984). The available resolution permitted measurement of profiles only for the macroparticles, not for the microparticles. In Figure 6.8 sulfonic-group concentration profiles measured as a function of time during sulfonation of type B particles are shown. The profile in the macropores is indeed uniform, and concentration increases with time as expected. Although the microparticle profiles cannot be measured, results reported by Ahn et al. (1988) under similar conditions show that gelular resins are not swollen and that sulfonation occurs only at the external surface. Finally, Figure 6.9 confirms the sulfonic groups' concentration profiles indicated earlier for type C and D particles, respectively (see Table 6.1).

6.5.2   Applications to Reacting Systems

The effects of nonuniform distribution of sulfonic groups on the performance of resin particles have been demonstrated for several systems. Klein et al. (1984)

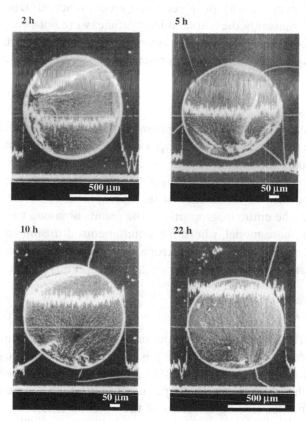

Figure 6.8. Scanning electron micrographs and EDX line scans of the cross section of resin particles sulfonated with H_2SO_4/CH_3NO_2 after 2, 5, 10, and 22 h. (From Klein et al., 1984.)

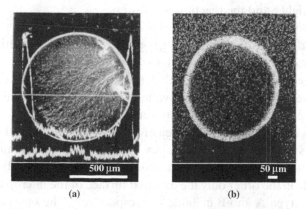

Figure 6.9. Scanning electron micrographs and EDX line scans of the cross section of a resin particle sulfonated with (a) $H_2SO_4/C_2H_4Cl_2$ and (b) SO_3 in a bed fluidized with N_2. (From Klein et al., 1984.)

used the *alkylation* of benzene with propylene as a model reaction. Due to the nonpolar nature of the reactants, the gelular microparticles were not swollen, and therefore the acid groups located inside the microparticles were not accessible to the reacting liquid. This explains the higher activity observed for the catalyst of type C in Table 6.1, as compared to uniform particles A, as well as to particles of type D, in which the largest fraction of acid groups are inside the gelular phase and are hence inaccessible.

Ahn et al. (1988) considered the *isomerization* of 1-butene as a model reaction to investigate the performance of three types of macroporous resin particles. Two of these, obtained by sulfonation of preswollen resin particles, were uniform but with different total loading. The third was of type B in Table 6.1, i.e., it contained sulfonic groups only on the surface of the microparticles, but uniformly distributed throughout the entire macroparticle. The results obtained were interpreted through a two-phase model, where the simultaneous diffusion and reaction processes were described at both the microparticle and macroparticle levels. In particular, the overall effectiveness factor of the macroporous particle, η was computed from

$$\eta = \eta_\alpha [\beta + (1 - \beta)\eta_\iota] \tag{6.6}$$

where η_α and η_ι are effectiveness factors accounting for diffusion in the macropores and the microparticle, respectively, while β is the fraction of sulfonic groups on the external surface of the microparticles, which are accessible immediately through diffusion in the macropores. The experiments were found to be consistent with the model results, leading to values of the microparticle effectiveness factor in the range 0.1–0.3 and to a reaction order with respect to sulfonic-group concentration equal to 2.4. The latter was explained by considering that the rate-determining step of the isomerization involved 2–3 sulfonic groups, in agreement with the results reported by Gates et al. (1972) for *t*-butanol dehydration. Catalyst B showed the best performance, since for a given loading of acid groups the local concentrations were higher and the reaction was faster.

Chee and Ihm (1986) investigated the effect of nonuniform distribution of active sites on catalyst *deactivation* during the gas-phase ethanol dehydration. In this system the produced water molecules interact with pairs of sulfonic groups, either forming hydrogen bonds or reacting with them to cause resin desulfonation. Both mechanisms were considered in a deactivation kinetic expression which was second-order with respect to sulfonic-group concentration. The two-phase model mentioned above was used to simulate the deactivation process observed experimentally in two types of sulfonated membranes. The macroporous membranes utilized had the same total loading of sulfonic groups but different distributions: one was uniform, while in the other only the external surface of the microparticles was functionalized, i.e. type A and B in Table 6.1, respectively. The observed deactivation patterns were quite different, and the nonuniform membrane exhibited the fastest deactivation rate due to the high surface concentration of acid groups. The experimental observations were in qualitative agreement with the two-phase model.

Nonuniform sulfonic-group distributions can also be employed to improve the *selectivity* in a multiple reaction system, by favoring the desired reaction relative to the undesired ones. Widdecke et al. (1986) studied the ether cleavage reaction

$$C_4H_9OCH_3 \rightarrow C_4H_8 + CH_3OH \tag{6.7}$$

which produces isobutene and methanol. Undesired reactions are mainly methanol dehydration and isobutene dimerization,

$$2CH_3OH \rightarrow CH_3OCH_3 + H_2O \tag{6.8}$$

$$2C_4H_8 \rightarrow C_8H_{16} \tag{6.9}$$

which are both catalyzed by acids. When a uniform resin is employed, the nonpolar ether reactant does not enter the sulfonated gelular microparticles, and the main reaction is restricted to the surface active groups in the macropores. However, the consecutive reactions, and particularly reaction (6.8) involving a polar reactant, take place also in the microparticles. Accordingly, when using a resin sulfonated only on the external surface of the microparticles, i.e. type B in Table 6.1, the product of the main reaction diffuses out of the macropores, reaction (6.8) is practically eliminated, and nearly 100% selectivity to the desired product is achieved (Widdecke et al., 1986).

Similar selectivity issues are pertinent to *esterification* and *transesterification* reactions, where an ester is produced by reacting an alcohol with an acid or another ester, respectively. These are large families of reactions, relevant to various products of industrial interest (Lundquist, 1995). In order to achieve satisfactory reaction rates, high temperatures (60–120°C) must be employed, which unfortunately give rise to undesired reactions catalyzed by the sulfonic groups. As mentioned above, these include the production of undesired ethers by alcohol condensation. Additionally, in the case of secondary or tertiary alcohols, dehydration reactions may also occur, leading to olefinic by-products. When uniform resins are utilized, the polar alcohol molecules have a strong tendency to enter the sulfonated gelular microparticles, where both the above undesired reactions can take place. Hence, similarly to ether cleavage, selectivity is improved by confining the sulfonic groups to the external surface of the microparticles. There they are equally accessible by the alcohol as well as the other reactant, i.e. acid or ester for esterification and transesterification, respectively.

Gelosa et al. (1998) studied the esterification of phthalic anhydride with ethanol, using type A and B resins (see Table 6.1). The reaction system consists of the two consecutive reactions

phthalic anhydride + ethanol → monoethylphthalate

monoethylphthalate + ethanol ⇌ diethylphthalate + water.

The first reaction is rapid and noncatalytic, while the second requires acidic catalyst. In Figure 6.10, conversion as a function of time for three different catalytic systems is shown. Curve 1 corresponds to homogeneous catalysis, obtained using a 95% solution of sulfuric acid, while curves 2 and 3 correspond to two different

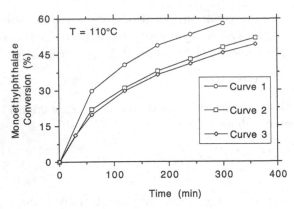

Figure 6.10. Conversion of monoethylphthalate as a function of time for three different catalytic systems: (1) sulfuric acid, (2) type A resin, (3) type B resin. (From Gelosa et al., 1998.)

types of sulfonated resin particles, i.e. uniform (type A in Table 6.1) and surface distribution (type B), respectively. The conversion profiles are similar as shown in Figure 6.10, but show significant differences in selectivity. In particular, for the homogeneous catalyst no by-products are formed, since the alcohol is contacted by the acid catalyst only in the presence of the monoethylphthalate, in which case only the esterification reaction is favored. However, when the uniform resin is utilized, ethanol accumulates in the gelular sulfonated microparticles, where the monoethylphthalate is almost completely excluded. This favors the ethanol condensation reaction, which leads to ~2.5% ether formation at the end of the reaction. This reaction is minimized with the surface-sulfonated resins, which lead to only 0.4% ether formation.

7

Preparation of Pellets with Nonuniform Distribution of Catalyst

onuniform catalyst distributions in porous supports are obtained primarily by multicomponent impregnation techniques. In general, an intermediate level of interaction between catalyst precursor and support is required, so that the precursor can attach to the support, but can also desorb if another competing adsorbing species is present. Depending upon the interplay between competitive adsorption and diffusion of the various species in the porous support, a variety of nonuniform catalyst distributions can be obtained. The above physicochemical processes are also encountered in chromatographic separations (Ruthven, 1984). This chapter is divided in two parts. The first deals with adsorption on powders, while the second is focused on simultaneous diffusion and adsorption phenomena.

Although diffusion–adsorption methods are dominant for the preparation of nonuniform catalyst pellets, other procedures have also been employed. One such technique is *deposition precipitation* in preshaped carriers (De Jong, 1991). It involves deposition inside pellets of insoluble compounds, such as hydroxides which are formed by a precipitation reaction. The latter can be induced by a change of solution pH. Immediately after imbibition, a pH profile develops inside the pellets, which depends on the initial solution pH and the isoelectric point of the carrier. Since precipitation reactions depend on pH, the insoluble compound distribution reflects the pH gradient. Hence, by appropriate choice of the impregnation conditions, precipitation can occur in either the inner (egg-yolk distribution) or the outer (eggshell distribution) region of the pellet. However, preparation of eggshell catalysts leads to the problem of precipitation outside the pellets. Therefore this method is especially suited for preparation of egg-yolk distributions, and has been utilized for preparation of Mo/SiO_2, $Cu/\gamma\text{-}Al_2O_3$, and $Ag/\gamma\text{-}Al_2O_3$ catalysts.

In other preparation techniques, the nonuniform catalyst pellet is obtained by assembling catalytic and noncatalytic layers. One such method is based on catalyst powder *granulation*. This can be realized by a two-stage fluidized/spouted bed (Scheuch et al., 1996). Catalyst powder premixed with a binder is deposited gradually on moist seeds (typically 0.5–2-mm diameter) of catalyst support, ultimately providing layered spherical particles. This method can in principle be extended for preparation of egg-white catalysts by using the corresponding eggshell catalysts as seed granules and depositing inert powder. Similarly, egg-yolk catalysts could be produced using uniform spherical catalyst particles as seed granules.

Coating is another technique that has been used for the preparation of egg-shell catalysts (Pernicone and Traina, 1984; Stiles and Koch, 1995). The coating equipment generally resembles pill-coating devices used in the pharmaceutical industry. The cores are coated in a rotating drum, with an appropriate catalytic *paint*. The thickness of the layer can be controlled by the amount of paint slurry introduced in the drum.

Tableting can also in principle be used to prepare egg-shell catalysts or other nonuniform distributions. In fact, specially designed presses are used in the pharmaceutical industry to prepare multilayer tablets. Examples of nonuniform pellet preparation by pressing together different layers are reported by Wu et al. (1990b) and Gavriilidis and Varma (1992).

Finally, *extrusion* can also be adapted for the preparation of nonuniform extrudates by appropriate design of the die.

7.1   Adsorption on Powders

Adsorption of metal complexes on oxide powders takes place without any diffusional limitation. The analysis can be based on phenomena which occur at the interface between the support and the impregnating solution, and has been developed largely in the colloid and interface science literature. Adsorption isotherm and surface ionization models have been used in order to quantitatively describe adsorption of catalyst precursors on oxide surfaces from aqueous solutions, with different outcomes depending on system and model complexity. The surface ionization models usually contain several parameters which can be adjusted to fit the experiments. As a consequence, different models can represent experimental data, with different values of the corresponding parameters. Adsorption isotherm models usually describe experiments within a limited range of conditions, but are popular because they are simpler to implement and require fewer parameter estimations.

In the following we first introduce adsorption isotherm models. Next, we discuss the effect of impregnation variables on adsorption, with reference to various experimental studies reported in the literature. Finally, we introduce surface ionization models, which are based on a detailed description of the solid–liquid interface.

7.1.1   Adsorption Isotherm Models

The most common relationship used to describe adsorption of catalyst precursors at constant pH is the Langmuir isotherm. In this case, the net rate of adsorption is given by

$$\frac{dn}{dt} = k^+ C(n_s - n) - k^- n \tag{7.1}$$

where n and C are the surface and fluid-phase precursor concentrations respectively, and n_s is the saturation capacity. At equilibrium, the amount adsorbed can be calculated by setting the transient term equal to zero, thus giving the Langmuir adsorption isotherm

$$\frac{n_e}{n_s} = \frac{KC_e}{1 + KC_e} \tag{7.2}$$

where n_e is the adsorbed amount, and C_e is the solution concentration, both at equilibrium. $K = k^+/k^-$ is the equilibrium adsorption constant, and a large value indicates strong adsorption. The Langmuir isotherm was first used to describe adsorption of gases on solids, but was later extended to describe adsorption from solutions. It assumes that the surface is energetically uniform, the energy of adsorption is independent of surface coverage, and the solution is dilute, so that the adsorbed species do not interact with each other. This simple equation represents well a broad spectrum of experimental adsorption data.

Adsorption of many metal precursors on various supports has been studied under equilibrium and transient conditions in batch systems. The most widely investigated system is H_2PtCl_6 on γ-Al_2O_3. A Langmuir model of adsorption was generally found to be suitable for fitting the experimental data (Santacesaria et al., 1977a; Shyr and Ernst, 1980; Castro et al., 1983; Jianguo et al., 1983; Heise and Schwarz, 1988; Xidong et al., 1988; Subramanian and Schwarz, 1991; Blachou and Philippopoulos, 1993; Papageorgiou et al., 1996). Figure 7.1 shows experimental results of H_2PtCl_6 adsorption on γ-Al_2O_3 under (a) equilibrium and (b) transient conditions. The surface saturation coverage n_s and adsorption equilibrium constant K can be determined from equilibrium measurements, and the adsorption rate constant k^+ from transient measurements. The surface saturation coverage obtained in Figure 7.1.a is $n_s = 1.55$ μmol/m^2, and the equilibrium adsorption

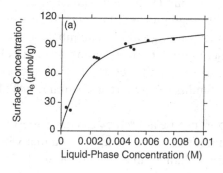

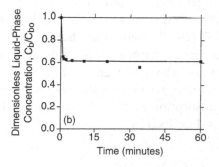

Figure 7.1. Adsorption of hexachloroplatinic acid on γ-Al_2O_3 powder: (a) equilibrium adsorption isotherm; (b) transient adsorption behavior ($C_{b0} = 0.01287$ M). The lines are the fitted Langmuir adsorption models, and the symbols denote experimental data. (From Papageorgiou et al., 1996.)

Table 7.1. Values of adsorption parameters of hexachloroplatinic acid on γ-Al$_2$O$_3$ in water solution.

Area (m^2/g)	K (l/mol)	k^+ (l/mol·s)	n_s (μmol/g)	n_s (μmol/m^2)	t_e (h)	Reference
177	459	1.21	265	1.50	3–8	Santacesaria et al., 1977a
150	1,330		110	0.73	2	Shyr and Ernst, 1980
170	31,000		140	0.82	6	Castro et al., 1983
245			275	1.12	6	Jianguo et al., 1983^a
	7,200	1.70				Scelza et al., 1986
190	1,550		151	0.79	0.3	Heise and Schwarz, 1988
195	110	0.18	290	1.48	2	Subramanian and Schwarz, 1991
227			205	0.90	1	Mang et al., 1993
78	606	2.46	121	1.55	1	Papageorgiou et al., 1996

^{a} Solid was η-Al$_2$O$_3$

constant is 606 l/mol. From Figure 7.1.b, the adsorption rate coefficient k^+ was calculated to be 2.46 l/mol·s. Values of these parameters reported in the literature range for surface saturation coverages from 0.73 to 1.55 μmol/m^2, and for the adsorption equilibrium constant from 110 to 31,000 l/mol (see Table 7.1). Adsorption characteristics of other systems that have been investigated include compounds of Pd on alumina (Sivaraj et al., 1991), Pd on silica and alumina (Schwarz et al., 1992), Pt on carbon (Hanika et al., 1982, 1983; Machek et al., 1983a,b), Rh on alumina (Hepburn et al., 1991a), Ni on alumina (Komiyama et al., 1980; Huang et al., 1986; Huang and Schwarz, 1987; Clause et al., 1992), Ni, Ba on alumina – one- and two-component impregnation – (Melo et al., 1980b), Cr on alumina (Chen and Anderson, 1973), Mo on alumina (Wang and Hall, 1980), Cr, Cu on alumina – one- and two-component impregnation – (Chen and Anderson, 1976), Cr, Mo, W on alumina – one-component impregnation – (Iannibello et al., 1979), Co, Mo on alumina – one- and two-component impregnation – (Iannibello and Mitchell, 1979; Cheng and Pereira, 1987; Hanika et al., 1987a,b; Sporka and Hanika, 1992), and organic acids on alumina (Kummert and Stumm, 1980; Engels et al., 1987).

The adsorption capacity of a support can be altered by calcination prior to impregnation (Schwarz, 1992). The number of surface sites on alumina which can be used for binding the metal precursor changes in a nonmonotonic fashion as a function of calcination temperature. In particular, a minimum in acidity and surface sites has been observed at about 600°C by some investigators (Tanabe, 1970; Sivaraj et al., 1991). Santacesaria et al. (1977b) found that the amount of platinum adsorbed during hexachloroplatinic acid impregnation on alumina samples calcined at different temperatures followed the same trend as the acidity.

In addition to the Langmuir isotherm, other models such as the Freundlich isotherm have been used for fitting adsorption data (cf. Benjamin and Leckie, 1981; Haworth, 1990). This isotherm is described by

$$n_e = KC_e^{1/m} \tag{7.3}$$

where K is a parameter related to the heat of adsorption, and m is a constant. Note that the amount of solute that can be adsorbed according to (7.3) is unlimited.

7.1.2   Effect of Impregnation Variables on Adsorption

There are various parameters that affect adsorption which simple models like the Langmuir isotherm do not take into account. These include solution pH, nature of support, surface heterogeneity, ionic strength, precursor speciation, presence of extraneous ions, and nature of solvent. When one or more of these parameters are changed, then even for the same precursor–support pair, rather different values of the Langmuir model parameters are obtained, as shown in Table 7.1 for the H_2PtCl_6/γ-Al_2O_3 system. Let us now review the effect of each of these parameters through various experimental evidence reported in the literature.

7.1.2.a Solution pH and Nature of Support

The solution pH can strongly affect adsorption in aqueous solutions. Santacesaria et al. (1977a) associated H_2PtCl_6 adsorption with acid attack on the alumina, dissolution of aluminum, subsequent formation of aluminum ions, and their final readsorption on the support after complexation with the hexachloroplatinate anions. Similar dissolution–reprecipitation of Al with Co, Ni, or Zn ions has also been observed even for nonaggressive pH, i.e. close to the isoelectric point of alumina (D'Espinose de la Caillerie et al., 1995). The amount of aluminum ions released into the solution depends on the crystallinity of the support: the solubility of alumina decreases as its crystallinity increases (Xidong et al., 1988; Subramanian et al., 1992). Various investigators who studied the effect of pH on chloroplatinic acid adsorption on alumina found that platinum adsorption shows a maximum at pH = 3–4 (Heise and Schwarz, 1985; Blachou and Philippopoulos 1993; Mang et al., 1993; Olsbye et al., 1997). Heise and Schwarz attributed the decrease of adsorbed platinum with decreasing pH to a decrease of adsorption sites due to alumina dissolution, and Blachou and Philippopoulos to the pH-dependent formation of platinum complexes with different affinity for the alumina surface.

Shah and Regalbuto (1994) argue that the reduction of platinum adsorption may be attributed entirely to the increased ionic strength of the solution at low pH, which leads to "double layer compression" or "electric screening", so that the equilibrium adsorption constant is effectively decreased. This means that oxide dissolution affects adsorption indirectly, through the ionic strength, and not directly by removing adsorption sites from the surface. In fact, Agashe and Regalbuto (1997) formulated a Langmuir model where adsorption equilibrium constants were calculated based on an overall Gibbs free energy comprising terms corresponding to coulombic attraction, repulsive solvation, and adjustable "chemical" interaction. This model satisfactorily predicted adsorption of metals as a function of pH, even without the use of the adjustable "chemical" term for certain systems, indicating that in those cases adsorption was physical (electrostatic) in nature.

Contescu and Vass (1987) studied the adsorption of chloro- and aminopalladium complexes on alumina and found that at constant pH the adsorption data followed the Langmuir isotherm. Both the adsorption capacity n_s and the adsorption constant K changed with pH and showed minima at the isoelectric point of alumina. For molybdate adsorption on alumina, it has been demonstrated that the adsorption sites are protonated hydroxyls, whose concentration can be decreased by increasing the pH (Spanos et al., 1990a,b; Goula et al., 1992b). Huang (1975) studied the adsorption of *ortho*-phosphate on hydrous γ-Al_2O_3 from dilute aqueous solution at constant ionic strength. Adsorption followed the Langmuir isotherm, except for large values of pH (>10). The adsorption constant K was found to vary with pH, and the adsorption capacity n_s showed a maximum at pH $= 4$, decreasing rapidly with increasing pH.

In all systems above, a decrease in anion adsorption was observed at pH larger than a certain value. This can be explained by invoking the simple electrostatic model of Brunelle (1978). According to this model, when oxide particles are suspended in aqueous solutions, a surface polarization results in net electrical surface charge. The type and magnitude of this charge are a function of the pH of the solution surrounding the particle and the nature of its surface. In general, in *acidic* solutions the surface is positively charged ($S-OH_2^+$), while in *basic* solutions the particles carry a negative surface charge ($S-O^-$). This is conveniently expressed by the following equilibrium reactions (Parfitt, 1976):

$$S-OH_2^+ \rightleftharpoons S-OH + H_s^+ \quad (K_{a1}) \tag{7.4a}$$
$\leftarrow$ Decreasing pH

$$S-OH \rightleftharpoons S-O^- + H_s^+ \quad (K_{a2}). \tag{7.4b}$$
Increasing pH $\rightarrow$

In between these two cases, a pH exists at which the net charge of the surface is zero. This value is characteristic of the oxide, and is called *point of zero charge* (PZC). Some authors use the terms point of zero charge (PZC) and *isoelectric point* (IEP) interchangeably. Properly, PZC is associated with a zero overall charge of the surface determined by potentiometric titration, and IEP with a zero electrophoretic mobility determined by microelectrophoresis (Parfitt, 1976). When the PZC and IEP points coincide (i.e. $pH_{PZC} = pH_{IEP}$), the pH_{PZC} is related to the two intrinsic acidity constants (7.4) by the following equation (Hohl and Stumm, 1976; Schwarz et al., 1992; Zhukov, 1996):

$$pH_{PZC} = \frac{pK_{a1} + pK_{a2}}{2}. \tag{7.5}$$

The protonated or deprotonated surface hydroxyl groups [see (7.4)] can be used to fix the metal ions to the support surface. In acidic solutions the positively charged surface will preferentially adsorb anions, while in basic solutions the negatively charged surface will adsorb cations (Brunelle, 1978; D'Aniello, 1981; Foger, 1984; Spanos et al., 1991). For example, for anionic adsorption the mechanism would be:

$$S-OH + H^+ \rightleftharpoons S-OH_2^+ \tag{7.6a}$$

$$y(S-OH_2^+) + M^{-n} \rightleftharpoons (S-OH_2^+)_y M^{-n}. \tag{7.6b}$$

In the absence of chemical reactions, the amounts of anion adsorbed can be controlled by adjusting the pH of the impregnating solution. Equivalently, a similar mechanism can be postulated for cationic adsorption in a basic solution:

$$S\text{–}OH + OH^- \rightleftharpoons S\text{–}O^- + H_2O \qquad (7.7a)$$

$$\text{n(S–O}^-) + M^{n+} \rightleftharpoons (S\text{–}O^-)_n M^{n+} \qquad (7.7b)$$

Adsorption of ions according to (7.6b) or (7.7b) may be simplistic, but has been demonstrated for several systems. However, more complicated mechanisms can be present, where adsorption of various anions produced during precursor speciation occurs, as for example during molybdena or chromia adsorption on alumina. In these cases, the metal is present in oxygen-containing anions, which not only adsorb on $S\text{–}OH_2^+$ but also react with S–OH sites of alumina (Lycourghiotis, 1995). Only amphoteric oxides such as aluminas, titanias, and chromia have the ability to adsorb both anions and cations. They are characterized by a pH_{IEP} in the range of 4–9 (Brunelle, 1978). Therefore, for $pH < pH_{IEP}$ they adsorb anions, while for $pH > pH_{IEP}$ they adsorb only cations. Acidic oxides such as zeolites, silica–aluminas, and silicas have low pH_{IEP}, and hence they adsorb cations, while basic oxides such as magnesia and lanthana have large pH_{IEP}, and they adsorb anions. A comprehensive collection of isoelectric points of oxides and hydroxides can be found in Parks (1965). It is worth noting that isoelectric points of different samples of the same oxide may vary markedly. This has been attributed to factors such as impurities, surface crystallinity, dehydration, and aging.

The isoelectric point of oxide supports is strongly influenced by the presence of foreign ions on its surface (Jiratova, 1981; Vordonis et al., 1984, 1986a,b, 1990; Mulcahy et al., 1987; Akratopulu et al., 1988; Mieth et al., 1990). Thus doping with suitable foreign ions can be used to influence the amount of ion adsorption that occurs for a given solution pH. The PZC is also affected by temperature. Akratopulu et al. (1986) found that precise regulation of the $\gamma\text{-Al}_2O_3\ pH_{PZC}$ can be achieved by varying the impregnation temperature. In this way the adsorption capacity of the support can be altered (Spanos et al., 1990b; Lycourghiotis, 1995). Alternatively, the PZC can be changed by calcination prior to impregnation. Sivaraj et al. (1991) found that the pH_{PZC} of alumina increased gradually with calcination temperature.

The solution pH affects not only the adsorption equilibrium constant but also the kinetic constants of the Langmuir model. Adsorption of hexachloroplatinic acid on alumina was studied by Blachou and Philippopoulos (1993) at two different solution pH values, chosen in such a way that the same amount of platinum was adsorbed at equilibrium (pH = 2.8 and 4.5). Adsorption kinetics were faster for the low-pH solution. This is probably related to the increased rate of formation of $S\text{–}OH_2^+$ at higher solution acidity, according to (7.6a).

We have so far tacitly assumed that each precursor ion adsorbs on one oppositely charged binding site. However, in some instances the amount adsorbed is only a fraction of the available sites (Santhanam et al., 1994; Contescu et al., 1993a, 1995a). Santhanam et al. (1994) attributed this discrepancy to the large size of ions which adsorb with one or two hydration sheaths intact, whereas Contescu et al. (1993a, 1995a) attributed it to steric constraints, which force the geometry of adsorbing complexes to match the arrangement of charged surface sites.

7.1.2.b Surface Heterogeneity

The electrostatic model of Brunelle (1978), according to which adsorption of cations/anions takes place in solutions with pH higher/lower than the pH_{PZC}, agrees with experimental data for many precursor–support systems (Kim et al., 1989; Bonivardi and Baltanas, 1990; Mulcahy et al., 1990; Vordonis et al., 1990; Sivaraj et al., 1991; Clause et al., 1992; Karakonstantis et al., 1992; Spanos et al., 1992; Zaki et al., 1992; Spielbauer et al., 1993). However, certain systems do not follow this general rule. For example, adsorption of cations on alumina has been reported for $pH < pH_{PZC}$ (Hohl and Stumm, 1976; Komiyama et al., 1980; Huang et al., 1986; Chu et al., 1989; Vordonis et al., 1992; Subramanian et al., 1992). This behavior can be explained by adsorption if one takes into account surface heterogeneity, i.e. the existence of several distinct types of sites with varying affinities for adsorbate ions (Hiemstra et al., 1989a,b; Contescu et al., 1993a,b, 1995a,b; Abello et al., 1995). Wang and Hall (1982) suggested that the presence of different crystal planes of alumina, with different local isoelectric points, was responsible for the surface heterogeneity, manifested by the observed breaks of the loading curves for molybdate and tungstate ions as a function of pH. Benjamin and Leckie (1981) invoked surface-site heterogeneity to explain deviations from Langmuir behavior. They studied adsorption of Cd, Cu, Zn, and Pb on amorphous iron oxyhydroxide and obtained data for a wide range of adsorption loadings. At low loadings all types of sites were available in excess, and the metal ion exhibited Langmuir adsorption. At higher loadings, deviation from the Langmuir behavior was observed, associated with a decrease in availability of the strongest binding sites, leading to a decrease in the apparent adsorption equilibrium constant.

Hiemstra et al. (1989a,b) developed a *multisite* model for proton adsorption, assigning different proton affinity constants to different types of surface groups existing at the solid–solution interface. The sites that develop after proton adsorption or desorption were considered to be singly, doubly, and triply metal-coordinated $-OH_2$ and $-O$ surface groups. The protonation and deprotonation reactions of surface $-OH$ can be represented by

$$M_n\text{--}OH_2^{nv} \rightleftharpoons M_n\text{--}OH^{nv-1} + H^+ \qquad (K_{n,1}) \tag{7.8a}$$

$$M_n\text{--}OH^{nv-1} \rightleftharpoons M_n\text{--}O^{nv-2} + H^+ \qquad (K_{n,2}) \tag{7.8b}$$

where $n = 1, 2, 3$ is the number of metal cations coordinated with surface $-OH$, and v is the bond valence reaching the oxygen or hydroxyl. If one assumes the presence of only one type of reactive surface group and if nv equals 1, the reactions (7.8) simplify to the reactions (7.4). The intrinsic affinity constants $K_{n,i}$ of the various surface groups depend on the local configuration of the surface, i.e. the number of coordinating cations, the valence of the cation, and the coordination number of the central cations in the crystal structure. The difference between log $K_{n,1}$ and log $K_{n,2}$ for proton binding on groups with the same surface configuration is so high (about 14 log K units) that only one protonation reaction can take place for each surface group within the normal pH range (3–11). Therefore, several Langmuir isotherms can coexist, one for each surface site configuration, i.e. $n = 1, 2, 3$.

Schwarz and Contescu (Contescu et al., 1993a, b 1995b) pictured the oxide surface as consisting of oxo and hydroxo groups characterized by different proton affinities. Using potentiometric titration data, they developed a method for calculating the proton affinity distribution of binding sites. In this way, the heterogeneity of surface proton-binding sites was demonstrated. For the case of γ-Al$_2$O$_3$, it was proposed that four categories of surface sites contribute to proton binding and surface charge development between pH 3 and 11. They were identified with types I-a (terminal, bound to tetrahedral Al), I-b (terminal, bound to octahedral Al), II-a (bridging, bound to octahedral and tetrahedral Al), and III (triple-coordinated, bound to octahedral Al) of surface hydroxyls. Type III are the most acidic sites, while type I are the most basic ones. These surface groups react with solution protons, depending on their log K_i and the solution pH, as follows:

$$[(Al_{Oh})_3–O]^{-0.5} + H^+ \rightleftharpoons [(Al_{Oh})_3–OH]^{+0.5}, \qquad \log K_{III} \sim 2.5$$
$$[(Al_{Oh})–O–(Al_{Td})]^{-0.75} + H^+ \rightleftharpoons [(Al_{Oh})–OH–(Al_{Td})]^{+0.25}, \quad \log K_{II\text{-}a} \sim 4.1$$
$$[(Al_{Td})–OH]^{-0.25} + H^+ \rightleftharpoons [(Al_{Td})–OH_2]^{+0.75}, \qquad \log K_{I\text{-}a} \sim 6.8$$
$$[(Al_{Oh})–OH]^{-0.5} + H^+ \rightleftharpoons [(Al_{Oh})–OH_2]^{+0.5}, \qquad \log K_{I\text{-}b} \sim 9.8.$$

$$(7.9)$$

In Figure 7.2 the experimental proton adsorption isotherm of alumina is superimposed on the calculated proton affinity distribution, $f(\log K)$, and the contributions of the various binding sites are shown. Due to site heterogeneity, different sites are present on the solid surface that are able to interact with *either* positive *or* negative species from the impregnating solution, *at any solution pH*. Changing

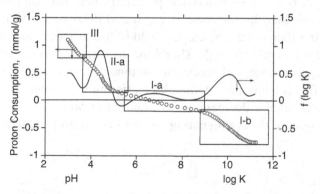

Figure 7.2. Proton binding isotherm (proton consumption vs pH) and the corresponding proton affinity distribution [$f(\log K)$ vs log K] for γ-Al$_2$O$_3$. Boxes assigned on basis of local coordination of OH groups. (From Contescu et al., 1995b.)

the pH results in variations in both number and type of surface sites with positive and negative charges. Depending on the solution pH, one or more types of surface sites may contribute to adsorption of the catalyst precursor. Hence adsorption equilibrium can be described by a combination of Langmuir isotherms, each associated with one type of surface site contributing to adsorption (Contescu et al., 1993a). This model predicts that even at the apparent PZC, some of the sites are still charged. This explains adsorption data which contradict Brunelle's model, such as the adsorption of cations at $pH < pH_{PZC}$, which in the case of γ-Al_2O_3 is attributed to a negative charge of II-a sites at low pH (Contescu et al., 1995b). In addition, geometrical constraints between the size and shape of the adsorbing ions and the ensembles of charged surface sites required for adsorption may also have to be allowed for, since these can be important factors that determine the amounts of catalyst precursor adsorbed (Contescu et al., 1993a, 1995a).

7.1.2.c Ionic Strength

The ionic strength of the impregnating solution affects the adsorption capacity of the support. It is defined by

$$I = \frac{1}{2} \sum Z_i^2 C_i \tag{7.10}$$

where Z_i is the valence and C_i the concentration of the individual ions. Ionic strength affects the electric double-layer thickness and activity coefficient of precursor ions in the impregnating solution (Heise and Schwarz, 1986). It has been demonstrated experimentally that the total uptake of metal precursor by the support decreases when the ionic strength of the impregnating solution increases (Heise and Schwarz, 1986, 1987; Shah and Regalbuto, 1994). This can be explained by considering that electrolyte ions modify the adsorption of metal precursor ions by altering the charge distribution near the surface of support. According to the Poisson–Boltzmann theory (Heise and Schwarz, 1987; Karpe and Ruckenstein, 1989), the surface charge creates an electrical potential whose extension into the bulk solution is determined by the ionic strength and decreases with increasing ionic strength. In other words, the electric field of the surface is increasingly shielded with increasing ionic strength. Therefore the attractive force between the oppositely charged surface and precursor ion diminishes, resulting in lower precursor uptake. This is confirmed by the observation that univalent and divalent compounds cause similar decreases in the amount of precursor adsorption when the ionic strengths of the impregnating solutions are equivalent (Heise and Schwarz, 1986).

7.1.2.d Precursor Speciation

The adsorption of metal complexes on the support can result from various types of precursor–support interactions which are not just of electrostatic nature. In general these include ion exchange or electrostatic adsorption (the immobilized species is not altered), ligand exchange with surface hydroxyl groups (the coordination

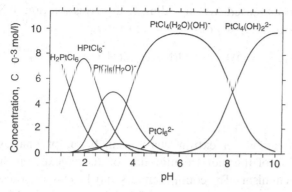

Figure 7.3. Concentration of platinum complexes in an aqueous solution of hexachloroplatinic acid as a function of pH, regulated by addition of aqueous solution of HCl or NaOH. (From Mang et al., 1993.)

sphere of the immobilized species is changed), and the formation of new chemical compounds at the interface (the chemical identity of the immobilized species is changed) (Westall, 1986; Schwarz et al., 1995). The interaction of metal precursor with the support depends on its actual form in solution, i.e. its chemical speciation. The latter is affected by the properties and concentration of the precursor, the solution pH, the ionic strength, and the presence of added or extraneous counterions (Schwarz, 1992). Metal ions in water are not bare, but hydrated. Thus they can participate in exchange reactions where the coordinated water molecules are exchanged for some preferred ligands. Hydrated multivalent metal ions can in principle donate a larger number of protons than that corresponding to their charge and can form anionic hydroxo–metal complexes.

For the case of aqueous solutions of chloroplatinic acid, Mang et al. (1993) calculated the pH-dependent distribution of chloro–aqua and chloro–hydroxo complexes of platinum(IV), when HCl and NaOH were used to adjust the pH. As shown in Figure 7.3, when the pH increases the charge of the dominant species changes from 0 to -2, while chlorine ligands are progressively replaced by H_2O and OH^-. It should be evident that the initial concentration of the precursor can influence the equilibrium concentrations of the formed complexes.

Palladium ions in water containing chloride and/or ammonium ions form various complex species with chloride and ammonium ligands, such as $PdCl_4^{2-}$, $PdCl_3$ $(H_2O)^-$, $Pd(NH_3)_4^{2+}$, with electric charges ranging from -2 to $+2$ depending on the solution pH (Contescu and Vass, 1987). Spielbauer et al. (1993) studied the adsorption of palladium amino–aqua complexes on γ-Al_2O_3 and SiO_2 by UV spectroscopy, and showed that the speciation of palladium ions takes place according to the following reactions, when pH adjustment was made by the use of ammonia:

$$[Pd(NH_3)_n(H_2O)_{4-n}]^{2+} + H_2O \rightleftharpoons [Pd(NH_3)_{n-1}(H_2O)_{4-(n-1)}]^{2+} + NH_3.$$

$$(7.11)$$

The charge of all complexes was $+2$, while the number of NH_3 ligands, n, increased from 0 to 4 as the pH was increased. When adding an aqueous solution of NaOH, amino–aqua and/or amino–hydroxo complexes are formed:

$$[Pd(NH_3)_{4-n}(H_2O)_n]^{2+} + OH^- \rightleftharpoons [Pd(NH_3)_{4-n}(H_2O)_{n-1}(OH)]^+ + H_2O.$$
(7.12)

7.1.2.e Coimpregnants

The presence of coimpregnants or extraneous ions in the impregnating solution can affect adsorption of the metal containing ions/complexes on the support through various mechanisms. The coimpregnants may be classified according to the mechanism through which they affect interfacial phenomena (Heise and Schwarz, 1987; Schwarz and Heise, 1990). The *first class* consists of inorganic salts such as NaCl, NaNO$_3$, and CaCl$_2$, which affect ionic strength. The *second class* includes inorganic acids and bases such as HCl, HNO$_3$, and NaOH, which affect the pH of the system. These compounds can also partially dissolve the oxide surface. The *third class* consists of compounds that compete with the metal ion for adsorption sites. Although many chemical species can adsorb on the surface, the strongest and most effective are those that contain hydroxyl, carboxyl, and phosphoryl groups, such as acetates, citrates, and phosphates. Note that these compounds also affect the pH and ionic strength of the solution.

Coimpregnants from the third class are commonly used to obtain nonuniform catalyst distributions. Single component adsorption parameters of citric acid on alumina are shown in Table 7.2. Large equilibrium adsorption constants have been obtained (2070–19,530 l/mol), indicating strong adsorption. Surface saturation coverage ranges from 1.43 to 3.35 μmol/m^2. Adsorption of other acids such as acetic, lactic, oxalic and tartaric on alumina were also found to exhibit Langmuir behavior (Engels et al., 1987; Jianguo et al., 1983).

During multicomponent adsorption, competition of the adsorbates for binding sites takes place. Jianguo et al. (1983) showed that the amount of H$_2$PtCl$_6$ adsorbed on η-Al$_2$O$_3$ was smaller in the presence of a coimpregnant. The decrease in amount adsorbed as compared with the single-component H$_2$PtCl$_6$ adsorption depended not only on the concentration but also on the type of coimpregnant. Adsorbed platinum in the presence of citric or tartaric acid was small, while in the presence

Table 7.2. Values of adsorption parameters of citric acid on γ-Al$_2$O$_3$ in water solution.

Area (m^2/g)	K (l/mol)	k^+ (l/mol · s)	n_s (μmol/g)	n_s (μmol/m^2)	t_e (h)	Reference
150			215	1.43	2	Shyr and Ernst, 1980
245			820	3.35	6	Jianguo et al., 1983[a]
220	2,070		444	2.02	2	Engels et al., 1987
78	19,530	7.97	209	2.68	1	Papageorgiou et al., 1996

[a] Solid was η-Al$_2$O$_3$.

of acetic, chloroacetic, or lactic acid it approached the values obtained by single-component H_2PtCl_6 adsorption. This is related to the fact that the first group of acids exhibited strong single-component adsorption, while for the second the adsorption was weak.

Competition for adsorption sites does not take place at low concentration of coadsorbing ions. Xue et al. (1988) investigated the adsorption on alumina of H_2PtCl_6 and H_2IrCl_6, which show similar adsorption characteristics. The adsorbed amount of a single metal from a bimetallic solution was the same as that from a pure solution. This is because at low concentrations, the spacing between adsorbed species on the surface is large and thus their interaction is negligible.

Papageorgiou et al. (1996), studied multicomponent adsorption of hexachloro-platinic (1) and citric (2) acids on γ-Al_2O_3. It was found that a *competitive Langmuir adsorption model* (i.e. no interactions between the adsorbates, either in the adsorbed or in the fluid phase)

$$\frac{dn_1}{dt} = k_1^+ C_1(n_s - n_1 - n_2) - k_1^- n_1 \tag{7.13a}$$

$$\frac{dn_2}{dt} = k_2^+ C_2(n_s - n_1 - n_2) - k_2^- n_2 \tag{7.13b}$$

was inconsistent with the measured data. Thus, an empirical model which took account of *solution effects* for hexachloroplatinic acid and *steric hinderance effects* for citric acid was formulated:

$$\frac{dn_1}{dt} = k_1^+ C_1(n_{s1} - n_1) - k_1^- n_1 - k_1^{sol} C_2 n_1 \tag{7.14a}$$

$$\frac{dn_2}{dt} = k_2^+ C_2(n_{s2} - n_2 - K_2^{st} n_1) - k_2^- n_2. \tag{7.14b}$$

The introduction of the solution-effects term was based on the observation that the amount of hexachloroplatinic acid desorbed when γ-Al_2O_3 was subsequently placed in a citric acid solution was in excess of that predicted by the single-component model but less than that predicted by the competitive Langmuir model. Also, the initial desorption rate of hexachloroplatinic acid was larger than that computed from both models, suggesting that its desorption is enhanced by the presence of citric acid in solution. The introduction of the term for steric hin-derance provided by the adsorbed platinum was based on the observation that the amount of citric acid adsorbed during multicomponent experiments was less than that expected from the competitive Langmuir model. The expressions for the multicomponent equilibrium isotherms including the new terms are

$$n_{1e} = \frac{K_1 C_{1e} n_{s1}}{1 + K_1 C_{1e} + K_1^{sol} C_{2e}} \tag{7.15a}$$

$$n_{2e} = \frac{K_2 C_{2e}}{1 + K_2 C_{2e}} (n_{s2} - K_2^{st} n_{1e}) \tag{7.15b}$$

where $K_1 = k_1^+/k_1^-$, $K_2 = k_2^+/k_2^-$, and $K_1^{sol} = k_1^{sol}/k_1^-$. The two new parameters, K_1^{sol} and K_2^{st}, were estimated by fitting the equilibrium data to be 735 l/mol and 3.18, respectively. For the multicomponent adsorption rate constants, it was found that the single-component value for hexachloroplatinic acid (2.46 l/mol·s) provided a satisfactory representation of the transient multicomponent data. However, the adsorption rate constant of citric acid had to be lowered to 0.2 l/mol·s in order to achieve the best fit.

7.1.2.f Nature of the Solvent

So far our discussion has been limited to aqueous systems, since water is the most common solvent used for catalyst impregnation. The solvent affects precursor adsorption, because in the case of electrostatic interaction the stability of the surface–ion pair depends on the dissociative power of the solvent. For solvents with large dielectric constants, this power is large and results in a lower stability of the surface–ion pair. Note that the dielectric constant of the solvent in the first layer above the surface can be smaller than its bulk value, as in the case of water, where the water dipoles in the first layer are aligned (Che and Bonneviot, 1988).

7.1.3 Surface Ionization Models

Surface ionization models characterize adsorption on oxide surfaces in terms of chemical and electrostatic interactions. They are typically based on the assumption that species bind to surface sites, and hence they are also called *surface complexation* models. Depending on the depiction of the solution–support interface and the species employed to describe the surface reactions, different models can be obtained. These models and their applicability have been discussed in various review papers (Westall and Hohl, 1980; Westall, 1986; Barrow and Bowden, 1987; Haworth, 1990).

When contacting an oxide with an aqueous solution, reactions of the type (7.4a,b) take place at the surface. They lead to the formation of protonated and deprotonated surface sites ($S–OH_2^+$ and $S–O^-$), which are responsible for the surface charge. In the absence of electrolyte or precursor ions, and neglecting the contributions of H^+ and OH^- (Westall, 1986; Charmas et al., 1995), the surface charge is given by

$$\sigma_0 = \frac{eN_A}{N_X}([S–OH_2^+] - [S–O^-]) \tag{7.16}$$

where σ_0 has units of C/m^2, N_A is Avogadro's number, N_X is the surface area of the oxide per unit of solution volume (m^2 of surface per m^3 of solution), and e is the elementary charge. In order to compute the concentration of surface sites, we consider the mass action equations corresponding to reactions (7.4a,b),

$$K_{a1} = \frac{[S–OH][H_s^+]}{[S–OH_2^+]} \tag{7.17}$$

$$K_{a2} = \frac{[S–O^-][H_s^+]}{[S–OH]}. \tag{7.18}$$

The concentration of surface species, $[H_s^+]$, is related to the bulk concentration $[H^+]$ by the Boltzmann distribution

$$[H_s^+] = [H^+] \exp(-e\Psi_0/k_B T) \tag{7.19}$$

where Ψ_0 is electrostatic potential at the surface. The exponential factor represents the electrostatic energy required to bring a charged species from the solution bulk to the adsorption surface. Using (7.19), equations (7.17) and (7.18) become

$$K_{a1} = \frac{[S-OH][H^+]}{[S-OH_2^+]} \exp(-e\Psi_0/k_B T) \tag{7.20}$$

$$K_{a2} = \frac{[S-O^-][H^+]}{[S-OH]} \exp(-e\Psi_0/k_B T). \tag{7.21}$$

The material balance equation for surface sites is

$$N_s = \frac{N_A}{N_X}([S-OH] + [S-O^-] + [S-OH_2^+]) \tag{7.22}$$

where N_s is the density of surface sites (sites/m^2). The equations above refer to the water/oxide system. When considering the adsorption of catalytic precursors, reactions of ions with surface sites must be included. This results in additional equations, i.e. mass action laws related to these new surface reactions, and in appropriate modifications of the equations above.

It is worth noting that the choice of the surface species is arbitrary and involves the introduction of additional parameters which have to be determined by fitting experimental data (Barrow and Bowden, 1987). A simple example is discussed later in the context of the triple-layer model.

The surface complexation model described by the four equations (7.16), (7.20)–(7.22) includes five unknowns ($[S-OH]$, $[S-O^-]$, $[S-OH_2^+]$, σ_0, Ψ_0) and three parameters (K_{a1}, K_{a2}, N_s), if the experimental pH is known. Thus, one more equation is required. This is given by a relation between surface charge σ_0 and surface potential Ψ_0, which is based on the electrostatic model employed to describe the interface. Some models for this purpose are discussed next.

7.1.3.a Constant-Capacitance Model

With regard to adsorption of ions on a charged surface, we can distinguish between *specific* and *nonspecific* adsorption. The first involves chemical bonds with the surface active sites, which are specific to the ions involved. The second is due to the electric potential of the surface and depends only on the ion charge. The *constant-capacitance*, or *single-layer*, model (see Figure 7.4) assumes that all specifically adsorbed ions are immobilized on the surface (plane 0) and contribute to the charge σ_0 together with the protonated and hydroxylated surface sites. All nonspecific adsorbed ions are excluded from the 0 plane. The potential is related linearly to surface charge by

$$\sigma_0 = c\Psi_0 \tag{7.23}$$

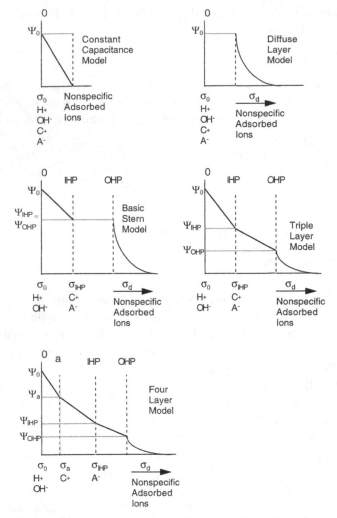

Figure 7.4. Electrostatic models for the surface–electrolyte-solution interface. In each case, adsorbent bulk is to the left of ordinate axis and solution to the right. The diagrams indicate the mean planes to which individual classes of ions are allocated along with the corresponding charges and show the change in electrostatic potential with distance. C$^+$ and A$^-$ refer to specifically adsorbed cations and anions respectively.

where c is the capacitance. The latter is an unknown parameter which has to be estimated by fitting experimental data. Its value is in principle valid only for the species and concentrations for which the data were obtained.

7.1.3.b Diffuse-Layer Model

The electrostatic properties of a surface in contact with an aqueous phase can be described by the diffuse-layer model. As shown in Figure 7.4, the interface is regarded as consisting of two regions: an inner region and an outer diffuse region (Gouy–Chapman layer). In the first, the ions are chemically bound to the surface and are relatively immobile. In the second, they are bound only by electrostatic interactions and undergo thermal motion. The distribution of ions in the diffuse region is given by the Poisson–Boltzmann equation with appropriate BCs. In particular, the following expression for the diffuse-layer charge is obtained in the case of univalent electrolyte:

$$\sigma_d = -(8\varepsilon_r\varepsilon_0 R_g TI)^{1/2}\sinh(F\Psi_0/2R_g T) \tag{7.24}$$

where ε_r is the relative permittivity of the solvent, ε_0 is the permittivity of vacuum, I is the solution ionic strength, and F is the Faraday constant. Due to electroneutrality, the surface charge is given by $\sigma_0 = -\sigma_d$. In the limit of low potential, the hyperbolic sine function can be approximated by its argument, leading to

$$\sigma_0 = (8\varepsilon_r\varepsilon_0 R_g TI)^{1/2}\frac{F}{2R_g T}\Psi_0. \tag{7.25}$$

The above expression is similar to equation (7.23) for the constant-capacitance model. The difference is that the capacitance is now determined by theory.

7.1.3.c Basic Stern Model

This is a version of the basic Stern model (1924) as implemented by Bowden et al. (1977). The protonated and deprotonated surface sites and the solvent ions H^+ and OH^- are assigned to the surface, where they contribute to the charge σ_0 and experience the potential Ψ_0. Specifically adsorbed ions are assigned to a plane, next to the surface, called the *inner Helmholz plane* (IHP), where they contribute to the charge σ_{IHP} and experience the potential Ψ_{IHP}. All nonspecifically adsorbed ions are excluded from the 0 plane and the IHP, and are assigned to the diffuse layer. As shown in Figure 7.4, the diffuse layer starts from a plane next to the IHP, which is called *outer Helmholz plane* (OHP). This is the location closest to the surface that the ions of the diffuse region can reach. The potentials of the IHP and OHP are set equal, $\Psi_{IHP} = \Psi_{OHP}$, and hence the capacitance between them is assumed to be infinite. The diffuse-layer charge σ_d can be calculated from (7.24) where Ψ_0 is replaced by Ψ_{OHP}. In the region between the 0 plane and the IHP, constant capacitance c_1 is assumed:

$$\sigma_0 = c_1(\Psi_0 - \Psi_{IHP}). \tag{7.26}$$

A charge balance equation for the IHP and an equation arising from electroneutrality are also required. These equations are discussed below in the context of the triple-layer model.

7.1.3.d Triple-Layer Model

The triple-layer model can be regarded as an extended Stern model (Davis et al., 1978; Davis and Leckie, 1978, 1980). It consists of two constant-capacitance layers and a diffuse layer. The protonated and deprotonated surface sites and the H^+ and OH^- ions are again located at the surface plane 0, and experience the potential Ψ_0. The IHP is separated from the surface by a region of constant capacitance c_1 and contains specifically adsorbed ions at potential Ψ_{IHP}. The IHP is separated from the OHP by a region of capacitance c_2. The potential at the OHP is Ψ_{OHP} and the total diffuse layer charge is σ_d. The relationships between charge and potential for the 0 plane, IHP, and OHP are

$$\sigma_0 = c_1(\Psi_0 - \Psi_{IHP}) \tag{7.27}$$

$$\sigma_{IHP} = c_2(\Psi_{IHP} - \Psi_{OHP}). \tag{7.28}$$

The charge in the diffuse layer is calculated by equation (7.24), where Ψ_0 is replaced by Ψ_{OHP}. The electroneutrality condition is given by

$$\sigma_0 + \sigma_{IHP} + \sigma_{OHP} = 0 \tag{7.29}$$

In a typical impregnating system for catalyst preparation, besides water and the solid, several electrolytes are present. The adsorption of electrolyte ions is considered to take place according to the site-binding model of Yates et al. (1974), i.e., the adsorbed ions form "ion pairs" with charged surface sites. Therefore, the electrolyte ions are considered as specifically adsorbed ions. The material and charge balance equations now have to include these additional species. For example, for a C^+A^- electrolyte, the material balance equation for surface sites is

$$N_s = \frac{N_A}{N_X}([S\text{–}OH] + [S\text{–}O^-] + [S\text{–}OH_2^+] + [S\text{–}O^-C^+] + [S\text{–}OH_2^+A^-]). \tag{7.30}$$

The charges at the surface and at the IHP are given by

$$\sigma_0 = \frac{eN_A}{N_X}([S\text{–}OH_2^+] + [S\text{–}OH_2^+A^-] - [S\text{–}O^-] - [S\text{–}O^-C^+]) \tag{7.31}$$

$$\sigma_{IHP} = \frac{eN_A}{N_X}([S\text{–}O^-C^+] - [S\text{–}OH_2^+A^-]). \tag{7.32}$$

In addition to mass action equations for the surface sites (7.20)–(7.21), mass action equations for the formation of surface complexes $S\text{–}O^-C^+$ and $S\text{–}OH_2^+A^-$ have to be included. If the surface reactions are described by

$$S\text{–}O^- + C_s^+ \rightleftharpoons S\text{–}O^-C^+ \qquad (K_C) \tag{7.33a}$$

$$S\text{–}OH_2^+ + A_s^- \rightleftharpoons S\text{–}OH_2^+A^- \qquad (K_A) \tag{7.33b}$$

(where subscript s denotes the surface), the mass action equations are

$$K_C = \frac{[S\text{–}O^-C^+]}{[S\text{–}O^-][C^+]\exp(-e\Psi_{IHP}/k_BT)} \tag{7.34}$$

$$K_A = \frac{[S\text{–}OH_2^+A^-]}{[S\text{–}OH_2^+][A^-]\exp(+e\Psi_{IHP}/k_BT)}. \tag{7.35}$$

The entire set of equations (7.20), (7.21), (7.24), (7.27)–(7.32), (7.34), (7.35) can be solved numerically at any given pH and electrolyte concentration. The parameters are K_{a1}, K_{a2}, K_A, K_C, N_s, N_X, c_1, and c_2, while the unknowns are [S–O$^-$C$^+$], [S–OH$_2^+$A$^-$], [S–OH], [S–O$^-$], [S–OH$_2^+$], σ_0, σ_{IHP}, σ_d, Ψ_0, Ψ_{IHP}, Ψ_{OHP}. Some of these parameters (N_s, N_X) can be determined by independent means: the density of reactive sites through surface titration (such as tritium exchange), and the surface area of the oxide by gas adsorption (Westall and Hohl, 1980). The other parameters are adjustable and can be estimated only through surface titration experiments. The model can incorporate more than one reaction to describe the adsorption of an ion on the surface, and a solute can give a large number of anions and cations according to its chemical speciation. Both lead to a larger number of adjustable parameters. Using the triple-layer model in its original form or with modifications, adsorption of various metal species has been described (Hachiya et al., 1984; Hayes and Leckie, 1986; Mang et al., 1993; Spielbauer et al., 1993). A revised version of this model has also been presented by Righetto et al. (1995).

7.1.3.e Four-Layer model

The four-layer model was introduced by Bowden and Barrow (Bowden et al., 1980; Bolan and Barrow, 1984; Barrow and Bowden, 1987) and can be regarded as an extension of the triple-layer model. A fourth plane, a, is introduced between the 0 plane and the IHP. The specifically adsorbed ions are located at the IHP or at the a plane, depending upon the strength of adsorption. The ions with stronger affinity to the surface are closer to it, i.e. at the a plane, while those with lower affinity are further away at the IHP. Charmas et al. (1995), in their implementation of the four-layer model, assigned cations to the a plane and anions to the IHP, because of their different size. The final equations are similar to those used for the triple-layer model, and are not discussed here in detail.

7.2 Simultaneous Diffusion and Adsorption in Pellets

Nonuniform catalyst profiles usually arise from the interaction of intrapellet flow, diffusion, and interfacial phenomena, depending upon the particular system and the impregnation process employed. In the latter, the support contacted with the impregnating solution is either dry (*dry impregnation*) or filled with the solvent (*wet impregnation*). In the first period of dry impregnation, i.e. *imbibition*, due to capillary forces, the solution fills the pores of the support and hence the solute is transported primarily by convective flow. After imbibition in dry impregnation, or during the entire course of wet impregnation, transport of solute is by diffusion. Adsorption of the solute by the pore walls occurs simultaneously with solute transport. The drying process following impregnation can also affect the final catalyst distribution. The total catalyst deposited on the internal surface of the pellet comprises the precursor adsorbed during impregnation and that precipitated from the unadsorbed solute during drying. The relative contributions of the two sources to local catalyst loading can be widely different, depending on the support and precursor properties, as well as on the catalyst preparation conditions.

The most common strategy for preparing nonuniformly distributed catalyst pellets is to first realize the desired distribution of the adsorbed precursor, and then dry the pellet, trying not to modify it. Therefore, in this section we focus on the impregnation process, both theoretically and experimentally, with the aim of determining the distribution of adsorbed precursor within the pellet. We conclude by discussing several studies which address nonuniform catalyst distributions from both experimental and modeling viewpoints.

7.2.1 Theoretical Studies

Various models of wet and dry impregnation have been developed (cf. Melo et al., 1980a; Lee and Aris, 1985). Intraparticle convective flow is due to capillary action, and its description is based on the Washburn equation or Darcy's law. Diffusion is typically described by Fick's law. Different variations of Langmuir isotherm models have been considered for the uptake of solute on the support, and a few investigators have also used surface ionization models.

7.2.1.a Dry Impregnation

For a porous spherical pellet the mass balances of impregnants during dry impregnation are

$$\varepsilon \frac{\partial C_{p,i}}{\partial t} + \rho_s \frac{\partial n_i}{\partial t} = \frac{\upsilon}{4\pi x^2} \frac{\partial C_{p,i}}{\partial x} + \frac{D_{e,i}}{x^2} \frac{\partial}{\partial x} \left(x^2 \frac{\partial C_{p,i}}{\partial x} \right) \tag{7.36}$$

with BCs

$$x = R: \qquad \upsilon(C_{b,i} - C_{p,i}) = 4\pi R^2 D_{e,i} \frac{\partial C_{p,i}}{\partial x} \tag{7.37a}$$

$$x = r_f: \qquad \rho_s \upsilon n_i = 4\pi r_f^2 \varepsilon D_{e,i} \frac{\partial C_{p,i}}{\partial x} \tag{7.37b}$$

and initial conditions

$$t = 0: \qquad C_{p,i}(x, 0) = n_i(x, 0) = 0 \tag{7.37c}$$

where $C_{p,i}$ is the concentration of the ith solute in the pores, ρ_s is the pellet density, and υ is the volumetric flow rate of imbibition, related to the position r_f of the advancing front by

$$\upsilon = -4\pi r_f^2 \varepsilon \frac{dr_f}{dt}. \tag{7.38}$$

The derivative $\partial n_i / \partial t$ represents solute uptake by the support and may be obtained from the Langmuir adsorption equation (7.1) (for single component adsorption) or (7.13) (for two-component competitive adsorption). In the case where adsorption is fast compared to the transport processes, the assumption of local adsorption equilibrium can be used. Accordingly, the concentrations in the adsorbed phase can be obtained directly from the concentrations in the pores through an adsorption model, such as for example the Langmuir isotherm equation (7.2). Therefore, by taking the adsorption isotherm in the general form

$$n_i = f_i(C_{p,1}, C_{p,2}, \ldots, C_{p,N}) \tag{7.39}$$

the expression for $\partial n_i / \partial t$ in equation (7.36) becomes

$$\frac{\partial n_i}{\partial t} = \sum_{k=1}^{N} \frac{\partial f_i}{\partial C_{p,k}} \frac{\partial C_{p,k}}{\partial t} \tag{7.40}$$

where N is the number of adsorbates.

In the imbibition stage of dry impregnation, the impregnating solution fills the pores of the support. Convective transport dominates over diffusion. The volumetric flow rate of imbibition, v, and the position r_f of the imbibed liquid at time t are computed through an appropriate hydrodynamic model. Two common approaches for this are based on Darcy's law (Lee and Aris, 1985) and on the Washburn equation (Vincent and Merrill, 1974). In particular, Lee and Aris (1985) derived equations for the front position and velocity under two different conditions: where the air occupying the pore space before penetration was either trapped in the pellet and compressed by the liquid front, or where it was released from the pellet. After the liquid front reaches the center of the pellet, i.e. $r_f = 0$, imbibition ceases and $v = 0$. From now on, diffusion is the only transport mechanism. In addition, depletion of solute in the external solution must be taken into account. The evolution of the system after imbibition is discussed later in the context of wet impregnation.

We now consider catalyst deposition during imbibition. In this period, diffusive transport may be neglected because convective transport dominates. For a *one-component* system, equations (7.36)–(7.37) can be simplified for the case of plug flow into a single pore (valid for slab shape pellets) to analyze the time-dependent flow of impregnating solution and dispersal of the impregnant (Vincent and Merrill, 1974):

$$\varepsilon \left(\frac{\partial C_{p,i}}{\partial t} + v \frac{\partial C_{p,i}}{\partial z} \right) = -a_p W_i \tag{7.41}$$

$$\rho_s \frac{\partial n_i}{\partial t} = a_p W_i \tag{7.42}$$

$$z = 0: \qquad C_{p,i}(0, t) = C_{b,i} \tag{7.43a}$$

$$t = 0: \qquad C_{p,i}(z, 0) = n_i(z, 0) = 0 \tag{7.43b}$$

where a_p is the pore surface area per unit volume, v is the velocity of the advancing front computed using the Washburn equation, and W_i is the mass flux of impregnants from liquid to the pore wall. Two situations were investigated for the solute mass flux W_i. In the first, the rate was controlled by mass transfer resistance across the liquid–solid interface, and in the second, by adsorption kinetics. Adsorption was described by the Langmuir model. Both cases were shown to yield similar eggshell impregnation profiles in systems where the uptake process was sufficiently fast. Moreover, in these cases it was possible to control the impregnation depth by adjusting the adsorption capacity.

For the same limiting cases, Lee (1984) developed analytical expressions for the catalyst distribution at the end of the imbibition period, considering both the precursor occluded in the pore and that adsorbed on the pore walls. This is the impregnation profile that would be obtained following imbibition with fast drying. An important extension included consideration of different one-dimensional pellet shapes (slab, cylinder, sphere) by allowing for the effect of geometry on the imbibition front velocity v in equation (7.41). For this purpose, a modification of the Washburn equation, developed by Komiyama et al. (1980), was used. It was found that the obtained nonuniform catalyst distributions degenerate to uniform ones only when sufficient time (of the order of days) is allowed for the equilibration between the occluded liquid and the adsorbed phase. Catalyst distribution in a pellet can also be influenced by adjusting imbibition conditions so that only partial penetration of the pores is achieved. Since convective transport of solution is due to capillary forces, critical parameters that affect the fraction of pore volume filled with liquid are solution viscosity, surface tension, support geometry, average pore size, and immersion time (Zhang and Schwarz, 1992).

Coimpregnation of pellets with catalyst precursor and coingredients can be used to manipulate intrapellet distribution of the active material. The model of Vincent and Merrill for imbibition was extended to a *multicomponent* system by Kulkarni et al. (1981). A competitive Langmuir adsorption model was considered. It was shown that when the equilibrium fractional coverage for a given concentration of precursor in the liquid phase is substantially smaller in the presence of a coimpregnant, it is possible for the precursor profile to exhibit a maximum within the pore. This condition is equivalent to stronger adsorption of the coimpregnant than of the catalyst precursor.

Analytical solutions for catalyst distribution at the end of the imbibition period were obtained by Lee (1985). The model was based on competitive Langmuir adsorption and the modified Washburn equation mentioned above for the front velocity. Since the advantage of utilizing a coimpregnant is best realized when adsorption is the controlling step, mass transfer resistance at the pore wall was neglected. Egg-white distributions were obtained if both impregnants adsorbed strongly, but the coimpregnant (species 2) adsorptivity was larger (i.e. K_1 was large and $K_2 > K_1$) and the rate of its adsorption was higher ($k_2^+ > k_1^+$) than that for the catalyst precursor. The catalyst layer's location and thickness can be manipulated by adjusting the bulk concentrations of the two impregnants. The location of the catalyst layer depends also on the pellet shape, because the latter affects the velocity of the penetrating liquid front. This occurs because the liquid front moves faster, as the pellet center is approached, in a spherical than in a cylindrical pellet; the velocity is independent of position in a slab pellet. Under identical preparation conditions, a slablike pellet has its shell closer to the center than a cylindrical or spherical pellet.

In an extensive analysis of catalyst impregnation, Lee and Aris (1985) considered various one- and two-component systems with both finite and infinite adsorption rates. The latter implies *instantaneous adsorption*, i.e., equilibrium prevails at any time and location throughout the wetted portion of the pellet. Under this condition, the concentration profiles take the form of shocks moving in time. For two-component impregnation, three regions, each with a constant composition, are found. The more strongly adsorbed coimpregnant (species 2) is present only in

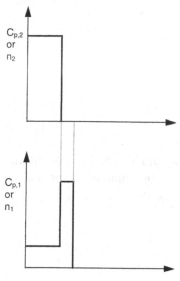

Figure 7.5. Profiles of pore fluid (C_p) and adsorbed (n) concentrations from two-component mixture imbibition with instantaneous adsorption. Component 2 is more strongly adsorbed. (From Lee and Aris, 1985.)

the outer region, as illustrated in Figure 7.5. A portion of the less strongly adsorbed precursor (species 1) passes through this region and penetrates deeper into the pellet, where it faces no competition for the adsorption sites. Thus, the surface concentration of adsorbed component 1 in this egg-white region is greater than that in the eggshell region. The highly concentrated component 1 adsorbed in the egg-white region is later washed off by the oncoming component 2, which causes the overshoot in the concentration profile of component 1. As time progresses, the band containing component 1 moves towards the interior of the pellet. When the center is reached, then the egg-yolk distribution is obtained.

In the case of *finite adsorption rate*, the sharp concentration profiles become blurred. Both the height and location of the catalyst layer are affected. As the relative rate of adsorption of the coimpregnant decreases, it becomes less effective in pushing the precursor into a narrow subsurface layer. However, if the rheological or interfacial properties of the impregnating solution are manipulated so as to slow down the liquid imbibition rate, nonuniformities in the catalyst distribution can be preserved.

Not only the adsorption rate but also the desorption rate affects the peak. If the desorption rate of the precursor is decreased, its adsorbed part is washed off slowly, smoothing the peak. Furthermore, the diffusion transport of solutes during imbibition, superimposed on finite rates of adsorption, intensifies the blurring of the otherwise sharp concentration profiles.

7.2.1.b Wet Impregnation

In wet impregnation the porous pellets, filled with pure solvent, are immersed in the impregnating solution. Therefore no convective solute transport takes place in the pores. Hence equation (7.36), with the convective flow term neglected, can be used to describe the mass balance of the solutes:

$$\varepsilon \frac{\partial C_{p,i}}{\partial t} + \rho_s \frac{\partial n_i}{\partial t} = \frac{D_{e,i}}{x^2} \frac{\partial}{\partial x} \left(x^2 \frac{\partial C_{p,i}}{\partial x} \right). \tag{7.44}$$

In addition, depletion of solutes in the external solution must be taken into account:

$$V_b \frac{dC_{b,i}}{dt} = -N_p D_{e,i} \, 4\pi R^2 \left(\frac{\partial C_{p,i}}{\partial x} \right)_{x=R} \tag{7.45}$$

where V_b is the volume of the external bulk fluid, $C_{b,i}$ is the concentration of the ith solute in the external bulk fluid, and N_p is the number of spherical pellets. If there are no external transport limitations, the BCs are

$$x = 0: \qquad \frac{\partial C_{p,i}}{\partial x} = 0 \tag{7.46a}$$

$$x = R: \qquad C_{p,i} = C_{b,i}. \tag{7.46b}$$

The initial conditions

$$t = 0: \qquad C_{b,i} = C_{b0,i}, \quad C_{p,i}(x, 0) = n_i(x, 0) = 0 \tag{7.47}$$

refer to coimpregnation, where precursor and coimpregnants are introduced at the beginning of the process. The same equations (7.44)–(7.46) can be used to describe the development of the catalyst distribution for dry impregnation after the imbibition process is complete. Different initial conditions for $C_{p,i}$ and n_i, however, have to be considered. These are given by the solution of the imbibition model discussed in the previous section.

Do (1985) studied one-component impregnation, where adsorption was described by the Langmuir isotherm. It was shown that when the impregnation is diffusion-controlled, a sharp eggshell solute distribution is obtained, particularly when there is no desorption of the active species, so that there is no chance for redistribution. The sharp front of the eggshell profile where the adsorption process takes place moves towards the pellet center as time increases. A higher adsorption capacity of the support results in a decrease of the front velocity and in an increase of the overall loading, but it does not have a significant effect on the form of the solute profile. When the solute's adsorption is slower, or its diffusivity is larger, it penetrates deeper into the support, yielding a smoother distribution front (Do, 1985; Vazquez et al., 1993). Similarly, in dry impregnation, the front of the eggshell profile obtained after imbibition is complete, moves towards the center of the pellet with increasing time (Lee and Aris, 1985). For slow adsorption, imbibition can yield a degenerate eggshell distribution with smoothly decreasing profile towards the pellet center. By either prolonging the impregnation time after imbibition or using wet impregnation with careful control of the impregnation time, a sharper eggshell distribution can be produced. This occurs because diffusive transport is much slower than convective transport. For the same reason, dry impregnation yields a thicker catalyst shell than wet impregnation, but this difference becomes less noticeable with longer impregnation time.

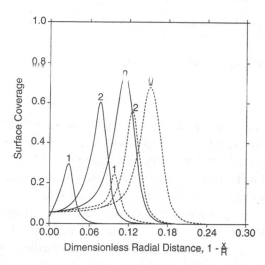

Figure 7.6. Comparison of adsorbed solute distribution between wet impregnation (solid lines) and dry impregnation (dashed lines): profile 1, 1 min; profile 2, 5 min; and profile 3, 10 min. Impregnation conditions $C_{b0,1} = 0.91 \times 10^{-2}$ M, $C_{b0,2} = 10C_{b0,1}$, $n_s = 1.88 \times 10^{-6}$ mol/m², $K_1 = 1.83 \times 10^3$ mol/l, $K_2 = 1.5K_1$, $k_1^+ = 21.11$ l/mol·s, $k_2^+ = 1.5k_1^+$, $D_{e,1} = 10^{-5}$ cm²/s, $D_{e,2} = 0.1D_{e,1}$. (From Lee and Aris, 1985.)

In the earlier part of this book, we have seen many instances where egg-white catalysts provide optimal performance. It is useful to describe here the general conditions which favor the formation of such distributions. The egg-white distribution obtained by two-component imbibition, as shown in Figure 7.5, is the result of competition between adsorption and convective transport. Once imbibition is over, the catalyst deposition becomes controlled by the interaction between adsorption and diffusion. This may either destroy or enhance the egg-whiteness of the distribution. Lee and Aris (1985) investigated different scenarios. For example, if the two components have same diffusivity and they are both weakly adsorbable, then the egg-white characteristic is not pronounced, and is quickly wiped out. However, when the precursor has larger diffusivity, egg-whiteness is modified but not destroyed. If in addition, the adsorption of both components is sufficiently strong, egg-whiteness is enhanced as the impregnation time is extended. As in one-component impregnation, the difference between dry and wet impregnation diminishes as the impregnation time is prolonged, as illustrated in Figure 7.6. Stronger adsorption of the coimpregnant is critical in obtaining egg-white catalysts. It can effectively push the precursor into a narrow, sharp, and concentrated band even when both components have the same diffusivity. Larger diffusivity of the precursor makes the overshoot of the egg-white distribution broader and higher. The adsorption capacities of the impregnants can also affect the catalyst profile. If they are small, the catalyst distribution evolves quickly from egg-white to egg-yolk and then becomes uniform. During this evolution, the peak is never very pronounced.

Lee and Aris (1985) have summarized the conclusions of their extensive simulations in the form of quantitative criteria for obtaining distinct egg-white or egg-yolk catalysts, in the case of purely competitive Langmuir adsorption:

1. $K_1 \gtrsim 10^3$ l/mol, $K_2 > K_1$.
2. $k_1^+ \gtrsim 10$ l/mol·s, $k_2^+ \geq k_1^+$.
3. $k_1^- \gtrsim 10^{-2}$ s⁻¹.

4. $D_{e,1} \geq D_{e,2}$ if $K_2 \gg K_1$; $D_{e,1} > D_{e,2}$ if $K_2 = K_1$.

5. $\rho_s n_{s,2}/\varepsilon C_{b0,2} > 1$.

6. Careful control of impregnation time is required.

7.2.1.c Effects of Electrokinetic and Ionic Dissociation Phenomena

The models discussed in previous subsections do not explicitly allow for the influence of electrokinetic transport and ionic dissociation on catalyst distribution. Ruckenstein and Karpe (1989) (see also Karpe and Ruckenstein, 1989) conducted a detailed theoretical study of wet impregnation taking these phenomena into account. A surface ionization model (see section 7.1.3) was used for ion adsorption. Chloroplatinic acid adsorption on γ-alumina was considered, in the presence of NaI to adjust the ionic strength, HI and NaOH to vary the pH, and NaNO$_3$, NaCl, and sodium citrate as competing coimpregnants. The axial flux inside the pores was given by the Nernst–Planck equation, which combines Fickian diffusion with ionic migration. The radial potential distribution within the pores was approximated by the Poisson–Boltzmann equation. Ionic fluxes are influenced by electrical potential due to surface charge, which in turn is affected by ionic strength, pH and adsorbed ions of the impregnating solution.

The effects of pH and ionic strength of the impregnating solution, nature of the support, and coimpregnating species on the Pt distribution profile were examined within this framework. It was found that for one-component impregnation (i.e. no competing coimpregnant), the axial distribution decreases from the surface to the interior of the pellet. When ionic strength increases, this distribution decreases in both height and depth. This is a consequence of the decreased adsorptivity of Pt due to increased ionic strength of the solution. Increase of external pH first increases the extent of adsorption (and therefore the height and depth of the adsorbed platinum distribution), which then passes through a maximum at pH 2.7, and eventually decreases. A comparison of this model with the more common model based on Fickian diffusion and Langmuir adsorption (where both electrokinetic transport and ionic dissociation are neglected; see section 7.2.1.b) and with the case where only electrokinetic transport is neglected (i.e., Fickian diffusion is the sole transport mechanism) is shown in Figure 7.7. These distributions are obtained using sodium citrate as the competing agent. It is clear that predictions of the three models are quite different. Electrokinetic phenomena affect the transport processes, because electrical potential in the axial direction increases the flux of coions and decreases the flux of counterions. The diffusion of protons and hydroxyls from the impregnating solution, coupled with the protonation–deprotonation equilibria of hydroxyl groups of the surface, can give rise to a nonuniform pH profile inside the pore, which affects not only ionic migration but also ion adsorption.

7.2.1.d Effect of Drying Conditions

The drying process following impregnation can alter catalyst distribution obtained at the end of the impregnation step. Lee and Aris (1985) investigated catalyst

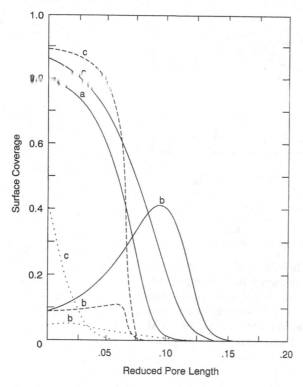

Figure 7.7. Model prediction of the effect of coimpregnation on the platinum distribution profile with citrate ion as coimpregnant. Solid curves: electrokinetic and ionic effects considered; dotted curves: electrokinetic effects neglected; dashed curves: electrokinetic and ionic dissociation effects neglected. Curve a: adsorbed impregnant in the absence of coimpregnation; curves b: adsorbed impregnant during coimpregnation; curves c: adsorbed coimpregnant. Chloroplatinic acid 0.5×10^{-3} M, sodium citrate 0.5×10^{-3} M, $pH_0 = 3.3$, time = 10 h, $I_0 = 10^{-3}$ M. (From Ruckenstein and Karpe, 1989.)

redistribution during drying, using a simplified model of a capillary tube where evaporation takes place at the open end of the tube. The liquid front moves from the pore interior to the external surface, as expected in the case of *slow* drying (see section 7.2.2.c). The model adopted is similar to that used for dry impregnation, i.e. equations (7.36)–(7.38), where now the front velocity has the opposite direction and is controlled by the drying rate. Solute precipitation and film flow were neglected, while several initial solute distributions were considered. They were altered during drying as a result of the interaction between adsorption and convective and diffusive transport. It was shown that drying has little effect on catalyst distribution in the case of strong adsorption. On the other hand, if adsorption is weak, drying has a smearing effect on eggshell and egg-white catalyst profiles developed at the end of the impregnation step. In this case, diffusion and

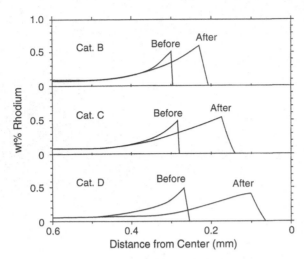

Figure 7.8. Model prediction of the effects of drying on the Rh impregnation profiles across the wall of γ-Al_2O_3 monoliths prepared by coimpregnation with $RhCl_3$ and HF. The curves are for adsorbed rhodium. Preparation conditions: 3.8×10^{-5} mol/cm^3 $RhCl_3$; Cat. B: 3.3 M HF; Cat. C: 6.6 M HF; Cat. D: 9.9 M HF. (From Hepburn et al., 1991a.)

flow bring the solutes back into a region of substantial adsorption near the pore mouth. However, the shape of the distribution can be preserved if the diffusivity of the precursor is larger than that of the coimpregnant, because the former moves inward faster and reaches available adsorption sites before the latter.

Hepburn et al. (1991a) studied catalyst redistribution during drying for the case where evaporation takes place at an interface receding towards the pellet interior, which is expected to occur in the case of *fast* drying (see section 7.2.2.c). The physicochemical parameters used were representative of γ-Al_2O_3 coimpregnated with $RhCl_3$ and HF, so as to realize an egg-white Rh distribution. The results indicated that drying causes a buildup of $RhCl_3$ and HF at the shrinking liquid front where evaporation occurs. As this front moves towards the center of the support, the increase in $RhCl_3$ and HF concentrations in the liquid causes a competition for unoccupied adsorption sites. $RhCl_3$ is pushed ahead of HF because the latter adsorbs more strongly than the former. As a result, after drying, the Rh band is displaced towards the interior and also becomes less sharp than the Rh profile before drying, as shown in Figure 7.8.

7.2.2   Experimental Studies

The catalyst precursor deposited on the surfaces of the pores of a pellet after impregnation is composed of two parts: the first is solute adsorbed on the support during the impregnation process, and the second is from precipitation of unadsorbed solute from the occluded pore liquid during the drying process. In this subsection, we discuss the role of both these processes in determining the final catalyst distribution through various experimental studies reported in the literature. These include single- and multicomponent impregnation as well as the effects of drying conditions.

7.2.2.a Single-Component Impregnation

When only the catalyst precursor is present during impregnation, and in the case where drying does not affect distribution of the catalytic material, the relative rates of transport and adsorption determine the distribution obtained. It has been observed experimentally for various systems that solute transport faster than adsorption yields relatively uniform distributions, while adsorption faster than transport yields eggshell distributions (Chen and Anderson, 1976; Summers and Ausen, 1978; Nicolescu et al., 1990). For intermediate situations, various types of distributions are obtained. If the adsorbed species can desorb at a reasonable rate, redistribution can occur by desorption and migration due to diffusion in the pores. This can ultimately give a uniform catalyst distribution (Maatman and Prater, 1957). Since the strength of adsorption is directly related to the adsorption rate ($K = k^+/k^-$), large desorption rates are typically characteristic of weak and not of strong adsorption. We now analyze the parameters that affect adsorption and transport rates, and hence also influence the catalyst distribution.

The *nature of the precursor* strongly affects the active-element adsorption characteristics. Summers and Ausen (1978) studied adsorption of various metal-ammonia–chloride complexes and demonstrated that they had different adsorption characteristics depending on the number of chlorine and ammonium ions. In the case of ammonium paratungstate, $(NH_4)_{10}W_{12}O_{41}$, a variety of species containing one, six, and twelve tungsten atoms can be present in aqueous solution. High tungsten salt concentration and low pH favor the existence of twelve-atom species, which are large and interact strongly with alumina. As a result, they tend to block the pores at the exterior of the pellet, resulting in eggshell profiles which persist even after long impregnation times (Maitra et al., 1986). Moreover, depending on whether the precursor is an acidic, basic, or neutral compound, its presence affects the solution pH and through this also the adsorption characteristics (Roth and Reichard, 1972; Chen and Anderson, 1973).

The *type of solvent* used in impregnation also plays an important role. For a given precursor–support system, adsorption isotherms, saturation coverage, and adsorption kinetics are different in aqueous and in organic impregnating solutions (Hanika et al., 1982; Machek et al., 1983a,b). The nature of the solvent can also affect the catalyst distribution through its volatility characteristics (Burch and Hayes, 1997). A volatile solvent can result in faster drying of the pellet after impregnation, which, as discussed below, can greatly affect the catalyst distribution. Impregnation without solvent has also been attempted. In particular, Soled et al. (1995) impregnated silica pellets with molten cobalt nitrate. The melt penetrated the pellets slowly due to its high viscosity, finally yielding eggshell catalysts.

Using *dry or wet impregnation* leads to different distributions. For long impregnation times this difference decreases, since the contribution of imbibition to the final distribution decreases. For weak adsorption, Cervello et al. (1976)

and Melo et al. (1980b) found that dry impregnation yields a catalyst with higher loading and more uniformity than wet impregnation. For $CuCl_2$ adsorption on γ-Al_2O_3, Ott and Baiker (1983) showed that the penetration depth obtained from dry impregnation was larger than from wet impregnation. Summers and Ausen (1978) observed similar trends, but the difference was less significant because the impregnation time was long.

The *concentration of impregnating solution* and the *impregnation time* determine the thickness of the shell in the case of strong adsorption. It has been observed for various systems that extended contact with the bulk impregnating solution allows the solute to diffuse into the support and enlarge the eggshell (Maatman and Prater, 1957; Harriott, 1969; Santacesaria et al., 1977c; Baiker and Holstein, 1983; Lee and Aris, 1985; Nicolescu et al., 1990). For $Ni(NO_3)_2$ and $Ba(NO_3)_2$ impregnation of alumina, which is characterized by weak adsorption, wet impregnation with short contact time, or low concentration of the impregnant, produces catalysts with decreasing concentration profile towards the pellet center. Increasing any of the above operating parameters smooths catalyst nonuniformities (Melo et al., 1980b). Goula et al. (1992b) studied the impregnation of ammonium heptamolybdate on γ-Al_2O_3. During dry impregnation they observed that with increasing precursor concentration, less uniform distributions were obtained. Since the opposite trend was expected for weak adsorption, this behavior was attributed to a pH effect. It was confirmed that by independently increasing the pH, keeping constant precursor concentration, more uniform distributions were obtained. In addition, the effect of impregnation time was investigated when imbibition was incomplete. It was found that longer impregnation times lead to smoother eggshell profiles when imbibition was incomplete, while further increase after imbibition was complete resulted in sharpening the eggshell distribution. This is because adsorption is faster than diffusion, thus favoring precursor adsorption at the pore mouth.

Goula et al. (1992b) did not observe any effect of the *impregnating solution volume* on the catalyst distribution profiles, which is reasonable for weakly adsorbing species, since no significant precursor depletion occurs in the bulk. For the case of strong adsorption, though, the volume of the solution would be expected to have a similar effect as its concentration, since it would affect the amount of precursor available for adsorption. Even for long impregnation times, since desorption is slow, the precursor is distributed nonuniformly if the amount in solution is less than that required to saturate, whereas if there is an adequate amount of solute, a uniform distribution can be obtained (Maatman, 1959).

The impregnating solution's *pH* and *ionic strength* have an indirect but significant effect on adsorption characteristics of the precursor. By altering the pH in the direction which favors adsorption of the impregnating species, sharper eggshell distributions can be obtained (Maitra et al., 1986; Heise and Schwarz, 1985; Goula et al., 1992b). For the system H_2PtCl_6/γ-Al_2O_3, when the ionic strength of the impregnating solution was increased by addition of simple inorganic salts, Heise and Schwarz (1986) found that the penetration depth of the eggshell distribution was not affected, but the adsorbed precursor concentration decreased and the

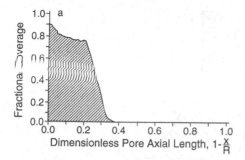

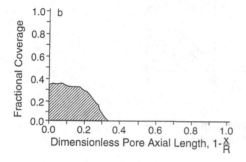

Figure 7.9. Experimental platinum adsorption profiles produced by impregnating γ-alumina pellets with 0.0025 M hexachloroplatinic acid and various sodium nitrate concentrations: (a) 0.0 M, (b) 0.005 M, and (c) 0.025 M. (From Heise and Schwarz, 1986.)

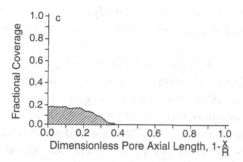

distribution front smoothed, as shown in Figure 7.9 for γ-Al$_2$O$_3$ impregnation with hexachloroplatinic acid in the presence of sodium citrate.

For ammonium heptamolybdate $[(NH_4)_6Mo_7O_{24}]$ impregnation of γ-Al$_2$O$_3$ extrudates, Goula et al. (1992b) smoothed eggshell catalyst profiles by decreasing the impregnation *temperature* and *doping* the support with fluoride ions. Using these methods, the pH$_{PZC}$ could be decreased and moved closer to the impregnation pH, thus reducing the concentration of adsorption sites.

7.2.2.b Multicomponent Impregnation

Most multicomponent impregnation studies concern platinum supported on γ-Al$_2$O$_3$. Shyr and Ernst (1980) studied the influence of a large number of acids and salts, used as coimpregnants with H$_2$PtCl$_6$, on the distribution of Pt within γ-Al$_2$O$_3$ pellets, and obtained the profiles shown in Figure 7.10. Among others, these include eggshell, egg-white, and egg-yolk distributions. The ability to prepare these distributions for Pt provides the motivation for investigating other

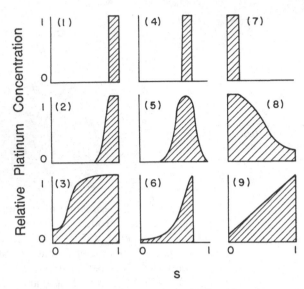

Figure 7.10. Types of Pt profiles obtained in coimpregnation experiments. (From Shyr and Ernst, 1980.)

multicomponent impregnation systems, in order to identify conditions for obtaining a desired catalyst distribution. The underlying principles have been discussed in section 7.2.1.

If HCl is used as *coimpregnant* along with chloroplatinic acid, a uniform platinum distribution is obtained (Maatman, 1959; Van den Berg and Rijnten, 1979; De Miguel et al., 1984; Scelza et al., 1986). When dibasic organic acids (for example citric or oxalic acid) are used as competitors, platinum is driven into the pellet interior, forming a subsurface layer with controlled location and width (Michalko, 1966b; Hoekstra, 1968; Summers and Hegedus, 1978; Becker and Nuttall, 1979; Li et al., 1985; Papageorgiou et al., 1996). HF has also been used as coimpregnant to prepare catalysts with subsurface layers of Pt (Hepburn et al., 1989a; Shyr and Ernst, 1980), Pd (Summers and Hegedus, 1978; Hepburn et al., 1989a), and Rh (Hegedus et al., 1979b; Hepburn et al., 1991a,c).

Hepburn et al. (1989c) prepared eggshell and egg-white Rh/γ-Al_2O_3 catalysts using hydrofluoric acid, hydrochloric acid, citric acid, and their sodium salts as coimpregnants. It was shown that rhodium can be driven beneath the external surface of the support more effectively by using coimpregnants with a low degree of dissociation in aqueous solution. As such, HF and citric acid result in egg-white distributions, while the other coimpregnants do not significantly influence the distribution type.

When adjusting the impregnating conditions to control the catalyst-layer location, the catalyst loading also usually changes. This occurs through a change of the width or local loading in the catalyst layer and/or the region between the external pellet surface and the catalyst layer (Hegedus et al., 1979b; Lee and Aris, 1985; Hepburn et al., 1991c; Papageorgiou et al., 1996). In catalyst preparation technology, it is desirable to control catalyst loading and location separately. This can be accomplished by using sequential impregnation as discussed below. For coimpregnation, a method that has been proposed to keep total loading constant is to form macromolecules which contain the metal and adsorb at the pellet exterior surface. This is done by introducing ethyl silicate in the impregnating solution so that $(-M-O-Si)_n$ structures are formed. Once a certain amount of macromolecules adheres over the pellet surface, no additional molecules adsorb. This results in a constant loading of metal ions in the catalyst pellet. Subsequently, the adsorbed macromolecules decompose on the pellet surface, and the metal ions diffuse into the pellet. In this case, the catalyst location is controlled by the impregnation time, and the catalyst loading by the concentration of precursor in the impregnating solution (Fujiyama et al., 1987; Otsuka et al., 1987).

In coimpregnation, two key parameters for controlling the catalyst distribution are impregnation time and competitor concentration. In general, it is observed that by increasing *impregnation time* the catalyst layer of an egg-white pellet is pushed deeper inward, yielding egg-yolk distributions. At long impregnation times, these subsurface cores can be washed out by back diffusion, producing uniform distributions with lower local loading (Becker and Nuttall, 1979; Hegedus et al., 1979b; Shyr and Ernst, 1980; Hepburn et al., 1991c; Papageorgiou et al., 1996). The effect of increasing *competitor concentration* is similar to that of increasing impregnation time: the catalyst layer is pushed deeper inside the pellet (Hepburn et al., 1989a; Papageorgiou et al., 1996). For Rh and Pt catalysts prepared using hydrofluoric acid as competitor, and Pt catalysts prepared using phosphoric acid as competitor, the local loading of the catalyst layer decreased as it was pushed inwards (Hepburn et al., 1989a; Schwarz and Heise, 1990). This is because these coimpregnants not only compete with catalyst precursor for active sites, but also alter the pH of the pore solution and hence local adsorption capacity of the support.

Another variable in coimpregnation is the *nature of the solvent*, which however has been only rarely investigated. For example, Alexiou and Sermon (1993) impregnated silica with $CuCl_2$ and $NiCl_2$ using methanol instead of water as solvent, and found that the metals penetrated deeper into the support.

Multicomponent coimpregnation can be performed starting with *dry or wet support*. The effect on catalyst loading is similar to that discussed earlier for single-component impregnation: higher total loadings are obtained by dry impregnation (Hepburn et al., 1991c; Melo et al., 1980b). Hepburn et al. (1991c) observed that the total loading of egg-white Rh/γ-Al_2O_3 catalysts increased with impregnation time in a manner depending on the initial state of support. For dry impregnation, the rhodium concentration increased mainly towards the external surface of the pellet, due to continued diffusion of rhodium precursor from the bulk liquid, after imbibition was complete. For wet impregnation, it increased uniformly in the region between the rhodium layer and the external surface. In addition, wet impregnation gave sharper catalyst layers than dry impregnation. This was attributed to faster adsorption and desorption rates relative to the transport rate for wet impregnation.

As discussed in section 7.2.1, the compounds used in multicomponent impregnation can be introduced simultaneously (*coimpregnation*) or successively (*sequential impregnation*) (Chen et al., 1980; Melo et al., 1980b; Papageorgiou et al., 1996). Papageorgiou et al. (1996) prepared Pt/γ-Al$_2$O$_3$ catalysts by impregnating the support with hexachloroplatinic acid and citric acid. When the pellet was impregnated with catalyst precursor and competitor sequentially, in that or the reverse order, similar egg-white catalyst distributions were obtained. The main difference was that if the metal precursor was impregnated first, the catalyst layer was thinner (Papageorgiou, 1984). Similar egg-white distributions were also obtained with coimpregnation, as discussed later in section 7.2.3. In coimpregnation experiments, platinum loading of the samples was affected not only by the initial hexachloroplatinic acid concentration, but also, to a smaller extent, by the citric acid concentration. The loading decreased with higher initial concentration of citric acid. However, during sequential impregnation, the platinum loading of the pellets depended primarily on the conditions of the first impregnation step. Approximately constant platinum loading was observed for pellets with the same first impregnation step, regardless of the citric acid concentration in the second step.

According to the classification scheme of Schwarz and Heise (1990) discussed in section 7.1.2.e, the above coimpregnants belong to the third class, because they compete with catalyst precursor for adsorption sites. However, other compounds belonging to the first or second class have also been used to influence the precursor distribution. As an example of the use of class 2 coimpregnants, Goula et al. (1992a) prepared eggshell and uniform molybdenum catalysts by heptamolybdate impregnation using acidic and alkaline solutions, respectively. In this case, the amount of Mo that can be deposited on γ-Al$_2$O$_3$ increases as the pH is decreased. By successive impregnation with NH$_4$OH solution, molybdena species adsorbed towards the pellet exterior were forced to desorb; hence eggshell and uniform distributions yielded egg-white and egg-yolk catalysts, respectively. Gaseous competitors have also been used to affect the precursor profile. Preadsorption of carbon dioxide on alumina has been found to promote uniform distribution of platinum, iridium, and rhenium (Kresge et al., 1992).

Multicomponent impregnation is also used for preparing *multimetallic* catalysts. Ardiles et al. (1986) and De Miguel et al. (1984) prepared Pt–Re/γ-Al$_2$O$_3$ catalysts by coimpregnation of H$_2$PtCl$_6$ and HReO$_4$. Eggshell catalysts were formed with layers of Pt and Re largely overlapping, with Pt more concentrated towards the pellet external surface. This is due to the higher affinity of chloroplatinic acid for the γ-Al$_2$O$_3$ support. If HCl was used as a coimpregnant, more uniform distributions of both metals were obtained. No significant differences were observed between distributions obtained with dry and with wet impregnation. On the other hand, Summers and Hegedus (1978) prepared Pt–Pd, Lyman et al. (1990) Pt–Rh, and Hegedus et al. (1979a) Pt–Pd–Rh catalysts, with separated layers of active materials, by successive impregnations with intermediate drying and calcination. Impregnation of the first metal precursor without competitor produces an eggshell layer, and subsequent impregnation of the second (and third) metal precursor along with a competitor deposits it in a subsurface band. When both metal precursors are impregnated simultaneously, they deposit in the same region when the support is initially dry, i.e., at the surface if no coimpregnant is used, and at a subsurface layer if a coimpregnant is present. If the support is prewetted, a slight separation of the catalyst peaks is observed (Lyman et al., 1990). Cheng and Pereira (1987) prepared cobalt–molybdenum catalysts by incipient wetness impregnation and found that the molybdenum profile was of the eggshell type, while the cobalt was distributed more uniformly. The penetration depth of molybdenum increased with decreasing surface area of the support Thus an interesting effect of the support surface area, which determines the number of active sites available for adsorption, was demonstrated.

7.2.2.c Effects of Drying

The drying step which follows the impregnation phase causes evaporation and flow of liquid in the pores, which can result in redistribution of impregnant in the pore volume. This can influence the adsorption–desorption equilibrium, leading to additional precursor adsorption, or affect solute precipitation from the pore-filling solution (Kheifets et al., 1979). The latter is the dominant mechanism of catalyst formation for weakly adsorbing precursors, but it can also be important for strongly adsorbing precursors if the initial concentration of the impregnating solution is greater than that required to saturate the support.

Drying of porous materials follows various stages characterized by drying rate (constant or falling), liquid-phase distribution (continuous or pendular), and predominant mechanism of moisture transport (capillary flow or vapor diffusion) (Berger and Pei, 1973; Neimark et al., 1981; Lee and Aris, 1985; Coulson and Richardson, 1991; Moyers and Baldwin, 1997). Two limiting regimes have been distinguished (see Figure 7.11): *fast* drying where vapor removal is much faster than capillary flow ($J_v \gg J_c$), and *slow* drying in the opposite case. In fast drying

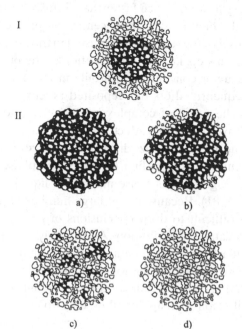

Figure 7.11. Regimes of drying. I, Fast drying. II, Slow drying: (a)–(d) indicate the various steps.

the evaporation front is continuous and recedes towards the pellet center as drying progresses, while in slow drying the evaporation front is initially located at the pellet surface, later portions of it move inward, and finally it breaks down so that isolated domains of liquid phase are formed (Neimark et al., 1976, 1981; Fenelonov et al., 1979). Evaporation at the pellet surface and at a moving front receding towards the pellet interior, depending on the drying conditions, have been experimentally observed (Hollewand and Gladden, 1992; Gladden, 1995). In addition, it has been shown that intrapellet moisture distribution can be quite different for pellets with different pore size distribution even when the overall drying rate is the same (Hollewand and Gladden, 1994).

The redistribution of solute is affected not only by liquid capillary flow but also by diffusive transport of the solute. The nonvolatile solute concentration increases in regions where the liquid vaporizes, thus creating concentration gradients, which cause diffusion of the solute (Lee and Aris, 1985). If it is fast enough as compared to vapor removal, diffusional relaxation smooths out concentration gradients. In the opposite case, high solute concentration exists at the evaporation front, and if it exceeds the saturation concentration, crystallites are formed. The interaction between mass transport and crystallite nucleation/growth can affect solute concentration gradients and catalyst dispersion, as discussed in section 2.4.1.

From the above discussion it follows that in general the solute distribution is determined by the relative magnitudes of evaporation rate, capillary flow, and solute diffusion. Since capillary flow is much faster than diffusive transport, for the purposes of precursor redistribution it is more appropriate to define as fast drying the regime where liquid evaporation is much faster than capillary flow ($J_v \gg J_c$), as slow drying the regime where liquid evaporation is much slower than diffusive transport ($J_d \gg J_v$), and as intermediate drying the regime where liquid evaporation is much slower than capillary flow but much faster than diffusive transport ($J_c \gg J_v \gg J_d$). With respect to the final precursor concentration profile in the pellet, fast drying allows one to preserve the same profile present in the pore-filling liquid at the end of the impregnation step, i.e., no redistribution of the precursor occurs. In all other cases the precursor concentration profile in the pore liquid changes during drying and consequently alters the deposited precursor profile in the pellet. The final profile is the result of complex interactions among the transport processes mentioned above, and is also affected by adsorption and precipitation in addition to the local and global pore size distribution and connectivity.

Other processes such as solvent adsorption, film flow (i.e. two-phase capillary flow) and recondensation through the gas phase may affect liquid transfer (Fenelonov, 1975; Neimark et al., 1981). Because of the large number of parameters and their interactions, it is difficult to draw conclusions of general validity about the effects of drying on catalyst distribution. Accordingly, conclusions reported in the literature, particularly in experimental work, are limited to the specific systems and conditions investigated. In the following we report some of these, to illustrate the effects of drying on catalyst distribution, although, owing to lack of quantitative data about diffusion and capillary flow, they do not provide a comprehensive picture.

For catalysts produced by nonadsorbable (or weakly adsorbed) impregnants, drying in the intermediate regime leads to an increase in the concentration of precursor near the external surface of the pellet, while in the slow regime diffusional relaxation smooths the concentration gradients. Van den Berg and Rijnten (1979) observed that $CuCl_2$ surface accumulation on γ-Al_2O_3 was smaller when dried slowly. For γ-Al_2O_3 pellets impregnated with ammonium heptamolybdate, the mildest drying conditions resulted in catalysts with more uniform Mo distribution and higher dispersion (Ochoa et al., 1979; Galiasso et al., 1983; Goula et al., 1992b). Harriott (1969) obtained a nearly uniform distribution of $AgNO_3$ with a low-surface-area alumina by drying slowly at 40°C. However, similar drying conditions yielded wide eggshell distribution for spheres soaked in silver lactate solution, indicating transition to the intermediate regime. Komiyama et al. (1980) found that $NiCl_2$ surface segregation can be suppressed by reaching the drying temperature of 110°C at 600°C/h, presumably owing to operation in the fast drying regime, whereas segregation occurred when drying at 100°C/h. On the other hand, for a similar system, Uemura et al. (1986, 1987) observed nickel segregation towards the outer surface of the support for high-severity drying conditions only.

In the case of strongly adsorbing precursors, single-component impregnation leads to eggshell distributions. Lee and Aris (1985) observed that these were similar regardless of the rate of drying. Chen et al. (1980) noticed that widening and smoothing of the eggshell layer of Cr/γ-Al_2O_3 and Cu/γ-Al_2O_3 catalysts occurred if they were left standing in air before drying at 150°C, and the effect was more pronounced for less strongly adsorbed solutes. Li et al. (1994) prepared Ni/Al_2O_3 catalysts, and even though uniform profiles were expected because of the prolonged wet impregnation, eggshell distributions were obtained due to redistribution during drying at 200°C. Santhanam et al. (1994) investigated the effect of adsorption strength by considering several precursors of Pt and Pd with different supports and impregnating conditions. They demonstrated that no solute migration occurred during drying of an initially homogeneously distributed precursor when strong interaction between solute and support existed, while surface segregation occurred for weaker interaction.

In all the above studies, the drying regime was influenced by adjusting the rate of liquid evaporation, for example through the drying temperature. Another option is to modify the capillary flow, for example by changing the viscosity of the impregnating solution. In particular, if the capillary flow of solvent is retarded, the solvent may not be supplied to the pellet surface at a fast enough rate to balance the evaporation rate. The liquid–air interface will then retreat towards the pellet center, until there is enough diffusional resistance between the interface and the pellet surface so that the evaporation rate is lowered to equal the capillary flow rate. Kotter and Riekert (1979, 1983b) investigated the effect of increasing the viscosity of the impregnating solution (by adding hydroxyethylcellulose) on the distribution of CuO on α-Al_2O_3. An eggshell profile was obtained, which extended progressively towards the pellet center with increasing viscosity, yielding ultimately a uniform distribution.

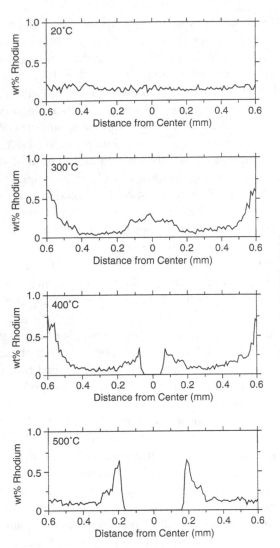

Figure 7.12. Experimentally obtained rhodium distribution across γ-Al_2O_3 slabs coimpregnated with 5.7×10^{-2} M $RhCl_3$ and 6.64 M HF, and dried at temperatures shown. (From Hepburn et al., 1989b.)

The effect of drying has also been studied in cases where at the end of impregnation an egg-white catalyst distribution is obtained. As expected, the active layer is preserved by fast drying, since the adsorbed precursor does not have time to redistribute (Hepburn et al., 1989b). Milder drying conditions can cause blurring of the catalyst layer, which becomes diffuse and wider and moves inwards (Lee and Aris, 1985). In addition, Hepburn et al. (1989b) and Becker and Nuttall (1979) noticed that in such cases, catalyst became concentrated also towards the external surface, giving rise to bimodal distributions, as shown in Figure 7.12. By drying slowly at 20°C, all peaks were completely washed out, giving a uniform catalyst (Hepburn et al., 1989b). In light of the discussion earlier in this section, it appears that drying at 20°C corresponds to slow, at 300 and 400°C to intermediate, and at 500°C to fast drying. Finally, two other alternatives for minimizing catalyst redistribution during drying are worth mentioning: immobilization of precursor before drying through liquid-phase reduction (Harriott, 1969; Becker and Nuttall, 1979; Kunimori et al., 1982), and freeze drying with liquid nitrogen (Chu et al., 1989).

7.2.2.d Determination of Catalyst Distribution

Measurements of local catalyst loading are required to establish the catalyst distribution. This is done by sectioning the catalyst pellet and then analyzing the internal catalyst concentration profile with a suitable technique. Although we do not describe these techniques, it is useful to list those that have been employed in the experimental studies reported in the literature. The most common technique is electron probe microanalysis (EPMA); other methods include light transmission, staining, spectroscopic analysis, and autoradiography. A summary of catalyst systems investigated using these methods is given in Table 7.3.

It is worth noting that measurement of the catalyst distribution, i.e. local loading as a function of position, does not necessarily mean that the behavior of the pellet under reaction conditions can be predicted, even for structure-insensitive reactions. The reason is that the reaction rate is proportional to the number of active sites, which is proportional to the catalytic surface area and not to the loading. The number and type of active sites are generally influenced by the conditions employed during impregnation. For example, it has been observed that adsorption can lead to species more difficult to reduce than those formed by precipitation from the pore-filling solution (Baiker and Holstein, 1983; Baiker et al., 1986; Goula et al., 1992a).

7.2.3 Comparison of Model Calculations with Experimental Studies

In order to improve our understanding of impregnation processes and to determine the reliability of mathematical models described in the previous section, it is useful to compare model predictions with experimental results. In the following, we discuss several comparisons that have been reported in the literature for different systems.

7.2.3.a Dry Impregnation

Komiyama et al. (1980) prepared nickel catalysts by dry impregnation of γ-Al_2O_3 with $NiCl_2$. Uniform and eggshell distributions were obtained by changing the initial concentration of the impregnating solution. Coimpregnation experiments utilizing HNO_3 resulted in egg-yolk distributions. This behavior was rationalized in terms of a pH effect. Since $NiCl_2$ does not adsorb below pH = 4.5, coimpregnation with HNO_3 inhibits Ni adsorption in the low-pH regions which are encountered towards the pellet surface. The experiments were simulated with the single-pore model (7.41), (7.42), which describes dry impregnation during the imbibition period. The precursor flux W_i from the impregnating solution to the pore wall was calculated assuming either mass transfer or adsorption kinetic control. Adsorption of $NiCl_2$ and HNO_3 was described by Langmuir and linear kinetics, respectively.

Table 7.3. Measurement methods of catalyst distribution and systems investigated.

Catalyst System	References
A. Electron Probe Microanalysis	
Co/Al_2O_3	Maitra et al., 1986; Otsuka et al., 1987
$Co-Mo/Al_2O_3$	Cheng and Pereira, 1987
Cr/Al_2O_3	Mathur et al., 1972; Chen and Anderson, 1973, 1976; Chen et al., 1980
Cu/Al_2O_3	Chen and Anderson, 1976; Chen et al., 1980; Van den Berg and Rijnten, 1979; Ott and Baiker, 1983; Baiker and Holstein, 1983; Baiker et al., 1986
Ir/Al_2O_3	Kresge et al., 1992
Mo/Al_2O_3	Ochoa et al., 1979; Galiasso et al., 1983; Duncombe and Weller, 1985; Goula et al., 1992a; Santhanam et al., 1994
Mo/SiO_2	Santhanam et al., 1994
Ni/Al_2O_3	Cervello et al., 1976; Komiyama et al., 1980; Maitra et al., 1986; Fujiyama et al., 1987; Chu et al., 1989; Uemura et al., 1989; Li et al., 1994
$Ni-Ba/Al_2O_3$	Melo et al., 1980b
$Ni-Mo/Al_2O_3$	Souza et al., 1989; Gazimzyanov et al., 1992
$Ni-P/Al_2O_3$	Ko and Chou, 1994
Pd/Al_2O_3	Satterfield et al., 1969; Hepburn et al., 1989a; Lin and Chou, 1994, 1995
Pt/Al_2O_3	Roth and Reichard, 1972; Becker and Nuttall, 1979; Van den Berg and Rijnten, 1979; Shyr and Ernst, 1980; Hepburn et al., 1989a; Kresge et al., 1992
$Pt-Bi/C$	Kimura, 1993
$Pt-Rh/Al_2O_3$	Lyman et al., 1990
Re/Al_2O_3	Kresge et al., 1992
Rh/Al_2O_3	Hepburn et al., 1989a,b,c, 1991a,b,c
Ru/Al_2O_3	Kempling and Anderson, 1970
W/Al_2O_3	Maitra et al., 1986
B. Light Transmission	
Ni/Al_2O_3	Komiyama et al., 1980
Pt/Al_2O_3	Heise and Schwarz, 1985, 1986, 1988; Schwarz and Heise, 1990
$Pt-Re/Al_2O_3$	Ardiles et al., 1986
C. Staining	
Mo/Al_2O_3	Srinivasan et al., 1979
Pt/Al_2O_3	Castro et al., 1983; Papageorgiou et al., 1996
D. Spectroscopic Analysis	
Mo/Al_2O_3	Blanco et al., 1987; Goula et al., 1992a,b
E. Autoradiography	
Ag/Al_2O_3	Harriott, 1969

Only the imbibition period was considered. The convective flow rate was evaluated using the Washburn equation, and diffusive transport was neglected. By fitting experimental catalyst profiles, they concluded that adsorption was controlled by kinetics and not by mass transfer. The model was consistent with observed nickel profiles, for both one- and two-component systems.

Heise and Schwarz (1988) prepared platinum catalysts by dry impregnation of γ-Al$_2$O$_3$ pellets with hexachloroplatinic acid. Adsorption equilibrium data were obtained independently and shown to follow a Langmuir isotherm. In this case also, the impregnation model (7.41), (7.42) was used, assuming adsorption kinetics to control the precursor flux from the impregnating solution to the pore wall. The adsorption kinetic parameters were evaluated by fitting platinum uptake during the impregnation experiments. The calculated platinum concentration profiles in the pellets were shown to be in qualitative agreement with the experimental eggshell profiles.

Platinum profiles obtained after coimpregnation of γ-Al$_2$O$_3$ with hexachloro-platinic and phosphoric acids were also simulated (Schwarz and Heise, 1990). For this, a multicomponent single-pore model for dry impregnation as discussed in section 7.2.1.a was used (Kulkarni et al., 1981). Phosphoric acid adsorption was assumed to follow Langmuir kinetics with an adsorption constant three times greater than that for the platinate ion. Qualitative agreement was obtained only after effects of the coimpregnant on solution-interface chemistry were incorporated into the model, viz., the influence of decreasing pH was simulated by decreasing the adsorption capacity of the support.

Hepburn et al. (1991a) impregnated dry γ-Al$_2$O$_3$ honeycombs with rhodium trichloride and hydrofluoric acid, and obtained egg-white Rh catalyst distributions. For both impregnants, single-component equilibrium adsorption data were measured and fitted with Langmuir adsorption isotherms. Multicomponent adsorption was described by the competitive Langmuir model [see equation (7.13)]. The catalyst preparation was simulated by considering separately the two steps: imbibition and drying. For the first, the single-pore model (7.41), (7.42) was used, assuming adsorption kinetic control; the second was simulated through a diffusion–adsorption model, similar to equation (7.44), with appropriate BCs to take account of solvent evaporation. Two adjustable parameters were used, viz. the adsorption rate constant of RhCl$_3$ and the effective diffusivity of HF. The two models were used sequentially to reproduce the Rh profile measured experimentally. Good quantitative agreement was obtained by tuning the two adjustable parameters, as shown in Figure 7.13. The prediction of the peak height was particularly sensitive to the chosen value of the RhCl$_3$ adsorption rate constant, while the peak location was primarily controlled by the surface saturation constant determined from the adsorption equilibrium data.

7.2.3.b Wet Impregnation

For one-component wet impregnation, the diffusion–adsorption model (7.44), (7.45) is generally used. Under the assumption that precursor adsorption is rapid

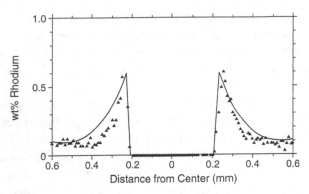

Figure 7.13. Comparison of model prediction with experimental results for Rh distribution across the wall of γ-Al_2O_3 monoliths prepared by coimpregnation with 3.8×10^{-5}-mol/cm^3 $RhCl_3$ and 3.3×10^{-3}-mol/cm^3 HF. (From Hepburn et al., 1991a.)

and irreversible, so that the time scale for adsorption is much smaller than that for diffusion, the model reduces to a shell progressive model. This has been used to simulate experimental results of $CuCl_2$ (Baiker and Holstein, 1983) and H_2PtCl_6 impregnation of γ-Al_2O_3 (Santacesaria et al., 1977c). In both cases, effective diffusivities were estimated by fitting the catalyst penetration depth as a function of time. Baiker and coworkers (Baiker and Holstein, 1983; Baiker et al., 1986) took into account both adsorbed copper and copper precipitated from the pore solution during drying, which ultimately gave rise to two types of copper species with different properties as identified by TPR. The agreement between experimental and calculated catalyst profiles was good, even though the model predicted a slower approach of the penetration front to the center of the pellet.

Hanika and coworkers (Hanika et al., 1982, 1983; Machek et al., 1983a) prepared Pt catalysts on activated carbon and alumina, utilizing aqueous and organic solutions of hexachloroplatinic acid, as well as phosphene and amine complexes of platinum. The one-component wet-impregnation model (7.44), (7.45) was used, assuming local adsorption equilibrium according to Langmuir or Freundlich isotherms. The effective diffusivities were calculated by fitting the time-dependent Pt uptake from the impregnating solution. The computed catalyst distributions were in qualitative agreement with the experimental data.

A two-component ($RhCl_3$–HF) wet-impregnation system was considered by Hegedus et al. (1979b) using the adsorption–diffusion model (7.44). The multicomponent adsorption was described by the competitive Langmuir model (7.13), whose parameters were estimated from single-component data of adsorption kinetics on powder support. In the Langmuir model, the various species were allowed to have different saturation concentrations. Diffusivities were determined by comparing the measurements of species uptake on pellets, using results of the diffusion–adsorption model. The model predicted Rh peak locations inside the γ-Al_2O_3 pellets satisfactorily.

A similar approach was followed by Scelza et al. (1986) to simulate wet coimpregnation of γ-Al$_2$O$_3$ pellets with H$_2$PtCl$_6$ and HCl. The same competitive Langmuir model was used to describe adsorption kinetics, but now assuming the same saturation concentration for all components. Moreover, the diffusion coefficients and the adsorption rate constants were estimated simultaneously by fitting experimental data on species uptake from the impregnating solution as a function of time. Unfortunately, different values of the adjustable parameters were obtained at each acid concentration, although the scatter was significant only for desorption rate constants. Modeling for various acid concentrations was carried out utilizing the corresponding set of diffusivities and adsorption parameters, and good agreement between theoretical and experimental profiles was obtained. It was observed that with increasing HCl concentration the profile develops from eggshell to uniform. The same approach was also used to simulate bimetallic wet coimpregnation. In particular, experimental data on H$_2$PtCl$_6$ and HReO$_4$ coimpregnation in the presence of HCl were considered (De Miguel et al., 1984; Ardiles et al., 1986). The predicted and experimental profiles of the various species after 12-h impregnation were in good agreement except the Re profiles, for which the predicted trend to form an egg-white distribution was not confirmed experimentally. Finally, it was found that the model was also able to predict Pt, Re, and Cl profiles obtained by dry impregnation, indicating that the imbibition period had no important influence due to the long impregnation time used in the experiments.

Melo et al. (1980b) prepared Ni and Ba catalysts on γ-Al$_2$O$_3$ by wet and dry impregnation. They performed single- and two-component experiments, where Ni(NO$_3$)$_2$ and Ba(NO$_3$)$_2$ were impregnated simultaneously or successively. In the latter case, Ni was impregnated first, and then, after calcination, Ba was impregnated as the second component. Wet impregnation was simulated with the adsorption–diffusion model (7.44), introducing external resistance to mass transfer and assuming local adsorption equilibrium inside the pellet. This was described through the multicomponent Langmuir model, whose parameters were estimated from independent equilibrium data (Melo et al., 1980a). The results of the wet-impregnation model were used as initial conditions for the drying model, to take proper account of the effect of drying on catalyst distribution. For dry impregnation a simplified model was used to describe the first imbibition period, and the wet-impregnation model was used again for the subsequent part of the process. The simplified model was based on the assumption that the penetration rate is much higher than the adsorption rate, so that adsorption could be neglected during imbibition (Melo et al., 1980b). From the first set of experiments involving one-species impregnation, it was found that diffusion and external mass transfer coefficients could not be evaluated independently but had to be estimated by fitting the species uptake data as a function of time. On the other hand, these parameter values allowed them to predict with reasonable accuracy the catalyst profiles obtained in two-species impregnation experiments. It is worth noting that to improve agreement with the experimental data, an adjustable parameter was introduced. This was the fraction of adsorption sites postulated to be accessible to nickel, while the remaining ones were available for both metals. On the whole, experimental profiles for one-component impregnation exhibited a weak eggshell shape, while the calculated ones were sharper. Two-component impregnations yielded more uniform distributions.

In the above studies, effective diffusivities were estimated by comparing experimental solute uptake measurements in pellets with results of a diffusion–adsorption model. In some cases, not only diffusivities but also other parameters (e.g. adsorption constants, external mass transfer coefficients) were adjusted to fit experimental data. Price and Varma (1988) developed an experimental procedure for the independent evaluation of diffusivity, based on the idea of decoupling adsorption and diffusion. This was done by running diffusion experiments at sufficiently high solute concentrations. In Langmuir adsorption, surface concentration of adsorbate does not change appreciably at saturation conditions, i.e. for fluid concentration values beyond a sufficiently high value. Thus, on keeping both the pore and the bulk fluid in this concentration range, adsorption and desorption do not occur. Hence, the diffusion–adsorption model equations (7.44), (7.45) are simplified, in that, $\partial n / \partial t = 0$. In order to ensure saturation conditions, pellets were placed in a high-concentration solution for several days, prior to their use in diffusivity experiments. They were then moved to a solution of different (but still sufficiently high) concentration. Effective diffusivities were determined by solving equations (7.44), (7.45) and calculating the value of D_e which gave the best fit to the time-dependent solute concentration change in the bulk solution.

Papageorgiou et al. (1996) prepared $Pt/\gamma\text{-}Al_2O_3$ pellets with a sharp internal step distribution of catalyst using coimpregnation and sequential impregnation of hexachloroplatinic and citric acids. In the sequential impregnation, the catalyst precursor was impregnated first, and then the competitive adsorbate. The latter adsorbed more strongly and pushed the former towards the center of the pellet, resulting in egg-white catalysts. Impregnation in the reverse order also produced similar egg-white catalysts, because the catalyst precursor moved past the pellet portion where citric acid had adsorbed (Papageorgiou, 1984). Similar behavior was observed for $\gamma\text{-}Al_2O_3$ pellets impregnated in a sequential fashion with chromium nitrate (strongly adsorbed) and copper nitrate (less strongly adsorbed) (Chen et al., 1980).

The cross sections of pellets prepared by sequential and coimpregnation techniques are shown in Figure 7.14. The dark rings in Figure 7.14.a–i are the platinum catalyst, while the lighter portion is the $\gamma\text{-}Al_2O_3$ support. For comparison, eggshell, egg-yolk, and uniform pellets are also shown in Figure 7.14.j–l. For both types of impregnations, the platinum band was located deeper inside the pellet for higher initial citric acid concentration (compare Figure 7.14.b–c, e–f, h–i). In addition, longer impregnation time forced the platinum to move closer to the pellet center, since the citric acid could diffuse further inside the pellet and displace the adsorbed platinum (compare Figure 7.14.a–b, d–e, g–h).

Increasing the initial hexachloroplatinic acid concentration did not have any noticeable effect on catalyst location; however, it resulted in catalysts with higher

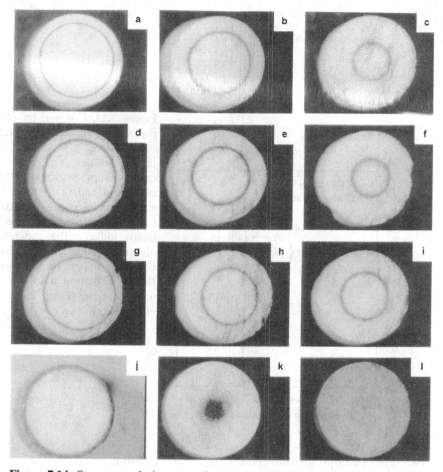

Figure 7.14. Step-type platinum catalysts prepared by sequential impregnation:
(a) 0.00589 M hexachloroplatinic acid (15 min), 0.1111 M citric acid (15 min),
(b) 0.00589 M hexachloroplatinic acid (15 min), 0.1111 M citric acid (60 min),
(c) 0.00589 M hexachloroplatinic acid (15 min), 0.3333 M citric acid (60 min), (d)
0.0117 M hexachloroplatinic acid (15 min), 0.1111 M citric acid (15 min), (e) 0.0117
M hexachloroplatinic acid (15 min), 0.1111 M citric acid (60 min), (f) 0.0117 M
hexachloroplatinic acid (15 min), 0.3333 M citric acid (60 min); and by coimpregnation: (g) 0.00589 M hexachloroplatinic acid and 0.1111 M citric acid (15 min), (h)
0.00589 M hexachloroplatinic acid and 0.1111 M citric acid (60 min), (i) 0.00589 M
hexachloroplatinic acid and 0.1667 M citric acid (60 min), (j) 0.0117 M hexachloroplatinic acid (15 min), (k) 0.0117 M hexachloroplatinic acid and 0.1667 M citric acid
(3 h), (l) 0.0117 M hexachloroplatinic acid and 0.1667 M citric acid (10 h). (From
Papageorgiou et al., 1996.)

loading and somewhat larger bandwidths (compare Figure 7.14.a–d, b–e, c–f).
The bandwidth was also influenced by the impregnation time, owing primarily to
geometric effects. Since the diameter of the ring where the platinum was located
decreased with time, its width increased to accommodate the deposited metal
(compare Figure 7.14.a–b, d–e, g–h). In general, for the same impregnation time
and adsorbate concentrations, similar penetration depths were achieved by the

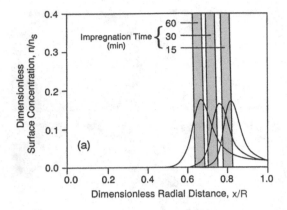

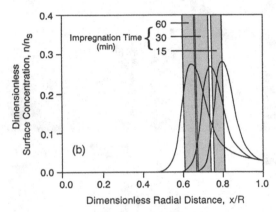

Figure 7.15. Comparison of model predictions and experimental measurements of the surface concentration profiles of platinum for sequential impregnation with 0.1111 M citric acid and (a) 0.00589 M hexachloroplatinic acid, (b) 0.0117 M hexachloroplatinic acid. The shaded bands indicate the experimentally measured platinum step positions, and the numbers on the bands give the impregnation time in the citric acid solution. (From Papageorgiou et al., 1996.)

two types of impregnations (compare Figure 7.14.a–g, b–h), but the bandwidth was smaller (as small as 50 μm) when using sequential impregnation.

The eggshell distribution (Figure 7.14.j) was prepared by impregnating the pellet in a solution of hexachloroplatinic acid alone. The egg-yolk distribution (Figure 7.14.k) could be obtained by both sequential impregnation and coimpregnation, provided that the impregnation periods were sufficiently long. The uniform distribution (Figure 7.14.l) was prepared using the coimpregnation method. Note that the difference in the preparation of pellets depicted in Figure 7.14.k and l is only in the impregnation duration. Thus, for short coimpregnation time an egg-white deposition is obtained, while for longer impregnation times egg-yolk and ultimately uniform distributions are attained.

Modeling of coimpregnation and sequential impregnation of pellets was carried out using equations (7.44), (7.45). Figures 7.15 and 7.16 show the predicted dimensionless platinum surface concentration profiles, along with the experimentally measured platinum band positions denoted by the shadowed regions for the sequential-impregnation and coimpregnation studies, respectively. The numbers on the bands give the impregnation time (for sequential impregnation it is the time in citric acid solution). The penetration depth of platinum predicted by the

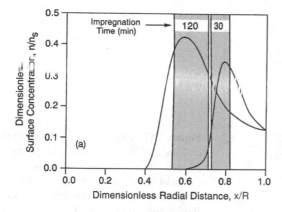

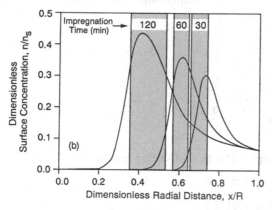

Figure 7.16. Comparison of model predictions and experimental measurements of the surface concentration profiles of platinum for coimpregnation with 0.0117 M hexachloroplatinic acid and (a) 0.0556 M citric acid, (b) 0.1111 M citric acid. The shaded bands indicate the experimentally measured platinum step positions, and the numbers on the bands give the impregnation time. (From Papageorgiou et al., 1996.)

model for both impregnation techniques, increases with increasing impregnation time, as also observed experimentally. In addition, the model is able to predict that the bandwidths are broader for coimpregnation than for sequential impregnation (compare Figures 7.15.b and 7.16.b), as well as the fact that the bandwidth increases with impregnation time. Note the model predicts that some adsorbed platinum extends from the peak to the pellet surface, with coimpregnation exhibiting more platinum in that region than sequential impregnation. This is consistent with experimental observations by various investigators (Kunimori et al., 1982; Schwarz and Heise, 1990; Hepburn et al., 1989a).

Figure 7.17 compares model predictions for the platinum peak location with experimental data, for various citric acid concentrations and impregnation times. The model agrees well with the data and demonstrates that on increasing the citric acid concentration for both impregnation techniques, the platinum band moves deeper within the pellet. The maximum difference between the experimental and theoretical peak locations is less than 8% of the pellet radius.

Chu et al. (1989) used the triple-layer electrostatic model discussed in section 7.1.3.d to simulate adsorption of $NiCl_2$ in the presence of HCl and NaCl on wet γ-Al_2O_3 pellets. As compared to Langmuir-based models, this provides a more realistic picture of surface chemistry, at the expense of model complexity.

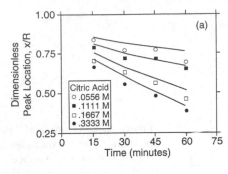

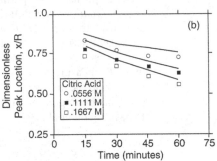

Figure 7.17. Experimental platinum peak locations and model predictions using 0.00589 M hexachloroplatinic acid and various citric acid concentrations: (a) sequential impregnation, and (b) coimpregnation. (From Papageorgiou et al., 1996.)

Adsorption of ions was described by the following reactions:

$$S{-}OH + H_2O + Ni^{2+} \rightleftharpoons S{-}O^-NiOH^+ + 2H^+ \tag{7.48}$$

$$S{-}OH + H^+ + Cl^- \rightleftharpoons S{-}OH_2^+Cl^- \tag{7.49}$$

$$S{-}OH + Na^+ \rightleftharpoons S{-}O^-Na^+ + H^+. \tag{7.50}$$

The support surface was assumed to be in equilibrium with the local pore concentrations of HCl and $NiCl_2$. The triple-layer model resulted in a system of twelve equations comprising charge–potential relationships, the electroneutrality condition, material balance for adsorption sites, charge balance equations, and mass action equations. The diffusion–adsorption equations (7.44) were used, but only for diffusion of H^+, OH^-, and Ni^{2+}, while the Na^+ and Cl^- pore concentrations were assumed to be constant throughout the pellet. The various model parameters were determined independently from adsorption measurements, and the effective diffusion coefficients were calculated from literature values for bulk diffusivities and a tortuosity factor of 2. Only the effective diffusivity of hydrogen ions was adjusted to obtain agreement with experimental results. An important feature of this model is the generation of protons upon nickel adsorption as described by equation (7.48), which explains the origin of the depression of Ni uptake under acidic conditions, also observed by Komiyama et al. (1980) for a $NiCl_2/HNO_3$ system as discussed in section 7.2.3.a. As nickel diffuses inside the pellet and adsorbs, a local excess of protons is generated, reducing nickel uptake and giving rise to a pH profile inside the pellet. This produces unique concave–convex metal profiles in the pellets, which were verified experimentally. These are maintained

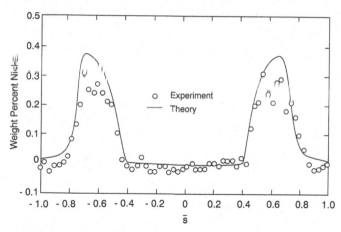

Figure 7.18. Nickel profile in a γ-alumina pellet after 26.75-h impregnation with 10^{-3} M NiCl$_2$ and 0.1 M NaCl, after acid pretreatment for 120 h. (From Chu et al., 1989.)

even after long impregnation times, due to the slow migration of protons relative to the metal ions. By controlling the initial pH profile, eggshell, egg-white, egg-yolk, and uniform catalysts can be obtained. For example, for producing egg-white catalysts, an acid pretreatment is required to yield a region of lower pH at the pellet surface than in the interior, followed by nickel impregnation. Nickel ions diffuse relatively quickly past the lower pH shell and adsorb as they reach the higher-pH internal region, creating an egg-white profile. For longer impregnation times, nickel diffusion continues and the egg-white region enlarges to produce an egg-yolk catalyst. Agreement of the model calculations with experimental nickel distributions was good in most cases, as shown in Figure 7.18.

Application of the Maximum Principle for Optimization of a Catalyst Distribution

The optimal catalyst distributions in a pellet and in an inert membrane reactor reported in sections 2.5.1 and Appendix B, respectively, have been obtained using the maximum principle. In this appendix we report the details of that analysis. In both cases three problems are addressed, which differ only in the objective function. They can all be reduced to the same form: determine the control function $\mu(s)$ which, under the constraints

$$\frac{d\mathbf{x}_1}{ds} = \mathbf{f}_1(\mathbf{x}_2) \tag{A.1}$$

$$\frac{d\mathbf{x}_2}{ds} = \mathbf{f}_2(\mathbf{x}_1, \mu) \tag{A.2}$$

$$\frac{d\mathbf{x}_3}{ds} = f_3(\mu) \tag{A.3}$$

$$\frac{d\mathbf{x}_4}{ds} = f_4(\mathbf{x}_1, \mu) \tag{A.4}$$

$$\frac{d\mathbf{x}_5}{ds} = f_5(\mathbf{x}_1, \mu) \tag{A.5}$$

with

$$\mathbf{x}_2 = s^n \frac{d\mathbf{x}_1}{ds} \tag{A.6}$$

and BCs

$$\mathbf{x}_2 = \underline{0} \qquad \text{at} \quad s = 0 \tag{A.7}$$

$$\mathbf{x}_2 = \mathbf{A}\,[\mathbf{f}(\mathbf{x}_2) - \mathbf{x}_1] \qquad \text{at} \quad s = 1 \tag{A.8}$$

$$x_3 = 0 \qquad \text{at} \quad s = 0 \tag{A.9}$$

$$x_3 = \frac{1}{n+1} \qquad \text{at} \quad s = 1 \tag{A.10}$$

$$x_4 = 0 \qquad \text{at} \quad s = 0 \tag{A.11}$$

$$x_5 = 0 \qquad \text{at} \quad s = 0 \tag{A.12}$$

maximizes the functional

$$I = G[\mathbf{x}_2|_1, x_4|_1, x_5|_1] + \int_0^1 R(\mathbf{x}_1, \mu)\, ds. \tag{A.13}$$

These equations represent both the pellet and the membrane problems through the following assignments:

Pellet problem.

Objective function	Effectiveness factor η	Selectivity S	Yield Y
x_1	x	x	x
x_2	γ	γ	γ
x_3	h	h	h
x_4	0	x_N	0
x_5	0	x_M	0
f_1	γs^{-n}	γs^{-n}	γs^{-n}
f_2	$\dfrac{\mu(s)}{1+B\mu(s)}\mathbf{w}(\mathbf{x})s^n$	$\dfrac{\mu(s)}{1+B\mu(s)}\mathbf{w}(\mathbf{x})s^n$	$\dfrac{\mu(s)}{1+B\mu(s)}\mathbf{w}(\mathbf{x})s^n$
f_3	$\mu(s)s^n$	$\mu(s)s^n$	$\mu(s)s^n$
f_4	0	$\dfrac{\mu(s)}{1+B\mu(s)}w_N(\mathbf{x})s^n$	0
f_5	0	$\dfrac{\mu(s)}{1+B\mu(s)}w_M(\mathbf{x})s^n$	0
f	x_f	x_f	x_f
I	η_M	S_{NM}	Y_{NM}
G	0	$\dfrac{D_M x_N(1)}{D_N x_M(1)}$	0
R	$\dfrac{\mu(s)}{1+B\mu(s)}w_M(\mathbf{x})s^n$	0	$\dfrac{\mu(s)}{1+B\mu(s)}w_N(\mathbf{x})s^n$

Membrane problem.

Objective function	Conversion X	Selectivity S	Purity P
x_1	x	x	x
x_2	γ	γ	γ
x_3	h	h	h
x_4	0	0	0
x_5	0	0	0
f_1	γs^{-n}	γs^{-n}	γs^{-n}
f_2	$\dfrac{\mu(s)}{1+B\mu(s)}\mathbf{w}(\mathbf{x})s^n$	$\dfrac{\mu(s)}{1+B\mu(s)}\mathbf{w}(\mathbf{x})s^n$	$\dfrac{\mu(s)}{1+B\mu(s)}\mathbf{w}(\mathbf{x})s^n$
f_3	$\mu(s)s^n$	$\mu(s)s^n$	$\mu(s)s^n$
f_4	0	0	0
f_5	0	0	0
f	$x_f = p_3[\gamma(1)]$	$x_f = p_3[\gamma(1)]$	$x_f = p_3[\gamma(1)]$
I	X_N	S_{DN}	P_{DN}
G	(B.27)	(B.30)	(B.32)
R	0	0	0

The optimization problem can be solved using the variational method described in more general terms in Ray (1981). Here we show its application to the cases of interest. Let us assume that $\bar{\mu}(s)$ is the optimal control function and express any other control function infinitesimally close to it as a perturbation about $\bar{\mu}(s)$:

$$\mu(s) = \bar{\mu}(s) + \delta\mu(s). \tag{A.14}$$

The state $\mathbf{x}(s)$ resulting from $\mu(s)$ can also be represented as a perturbation about the state $\bar{\mathbf{x}}(s)$ caused by the control function $\bar{\mu}(s)$:

$$\mathbf{x}(s) = \bar{\mathbf{x}}(s) + \delta\mathbf{x}(s). \tag{A.15}$$

Next we linearize equations (A.1)–(A.5) about the optimal function $\bar{\mu}$, to yield

$$\frac{d(\delta\mathbf{x}_1)}{ds} = \frac{\partial \mathbf{f}_1^T}{\partial \mathbf{x}_2} \delta\mathbf{x}_2 \tag{A.16}$$

$$\frac{d(\delta\mathbf{x}_2)}{ds} = \frac{\partial \mathbf{f}_2^T}{\partial \mathbf{x}_1} \delta\mathbf{x}_1 + \frac{\partial \mathbf{f}_2}{\partial \mu} \delta\mu \tag{A.17}$$

$$\frac{d(\delta x_3)}{ds} \doteq \frac{\partial f_3}{\partial \mu} \delta\mu \tag{A.18}$$

$$\frac{d(\delta x_4)}{ds} = \frac{\partial f_4}{\partial \mathbf{x}_1} \delta\mathbf{x}_1 + \frac{\partial f_4}{\partial \mu} \delta\mu \tag{A.19}$$

$$\frac{d(\delta x_5)}{ds} = \frac{\partial f_5}{\partial \mathbf{x}_1} \delta\mathbf{x}_1 + \frac{\partial f_5}{\partial \mu} \delta\mu. \tag{A.20}$$

Similarly, equation (A.13) gives

$$\delta I = \frac{\partial G}{\partial \mathbf{x}_2} \delta\mathbf{x}_2|_1 + \frac{\partial G}{\partial x_4} \delta x_4|_1 + \frac{\partial G}{\partial x_5} \delta x_5|_1 + \int_0^1 \left(\frac{\partial R}{\partial \mathbf{x}_1} \delta\mathbf{x}_1 + \frac{\partial R}{\partial \mu} \delta\mu \right) ds. \tag{A.21}$$

Let us now adjoin to the objective functional the linearized constraints [equations (A.16)–(A.20)] multiplied by the adjoint variables (Lagrange multipliers) $\underline{\lambda}_i(s)$:

$$\int_0^1 \lambda_1^T(s) \left(\frac{d(\delta\mathbf{x}_1)}{ds} - \frac{\partial \mathbf{f}_1^T}{\partial \mathbf{x}_2} \delta\mathbf{x}_2 \right) ds = 0 \tag{A.22}$$

$$\int_0^1 \lambda_2^T(s) \left(\frac{d(\delta\mathbf{x}_2)}{ds} - \frac{\partial \mathbf{f}_2^T}{\partial \mathbf{x}_1} \delta\mathbf{x}_1 - \frac{\partial \mathbf{f}_2}{\partial \mu} \delta\mu \right) ds = 0 \tag{A.23}$$

$$\int_0^1 \lambda_3(s) \left(\frac{d(\delta x_3)}{ds} - \frac{\partial f_3}{\partial \mu} \delta\mu \right) ds = 0 \tag{A.24}$$

$$\int_0^1 \lambda_4(s) \left(\frac{d(\delta x_4)}{ds} - \frac{\partial f_4}{\partial \mathbf{x}_1} \delta\mathbf{x}_1 - \frac{\partial f_4}{\partial \mu} \delta\mu \right) ds = 0 \tag{A.25}$$

$$\int_0^1 \lambda_5(s) \left(\frac{d(\delta x_5)}{ds} - \frac{\partial f_5}{\partial \mathbf{x}_1} \delta\mathbf{x}_1 - \frac{\partial f_5}{\partial \mu} \delta\mu \right) ds = 0. \tag{A.26}$$

If we require that equations (A.16)–(A.20) be satisfied everywhere, then the subtraction of equations (A.22)–(A.26) from equation (A.21) yields

$$\delta I = \frac{\partial G}{\partial \mathbf{x}_2} \delta \mathbf{x}_2|_1 + \frac{\partial G}{\partial x_4} \delta x_4|_1 + \frac{\partial G}{\partial x_5} \delta x_5|_1$$

$$+ \int_0^1 \left(\frac{\partial R}{\partial \mathbf{x}_1} + \boldsymbol{\lambda}_2^{\mathrm{T}} \frac{\partial \mathbf{f}_2^{\mathrm{T}}}{\partial \mathbf{x}_1} + \lambda_4 \frac{\partial f_4}{\partial \mathbf{x}_1} + \lambda_5 \frac{\partial f_5}{\partial \mathbf{x}_1} \right) \delta \mathbf{x}_1 \, ds + \int_0^1 \left(\boldsymbol{\lambda}_1^{\mathrm{T}} \frac{\partial \mathbf{f}_1^{\mathrm{T}}}{\partial \mathbf{x}_2} \right) \delta \mathbf{x}_2 \, ds$$

$$+ \int_0^1 \left(\frac{\partial R}{\partial \mu} + \boldsymbol{\lambda}_2^{\mathrm{T}} \frac{\partial \mathbf{f}_2}{\partial \mu} + \lambda_3 \frac{\partial f_3}{\partial \mu} + \lambda_4 \frac{\partial f_4}{\partial \mu} + \lambda_5 \frac{\partial f_5}{\partial \mu} \right) \delta \mu \, ds$$

$$- \int_0^1 \left(\boldsymbol{\lambda}_1^{\mathrm{T}} \frac{d(\delta \mathbf{x}_1)}{ds} + \boldsymbol{\lambda}_2^{\mathrm{T}} \frac{d(\delta \mathbf{x}_2)}{ds} + \lambda_3 \frac{d(\delta x_3)}{ds} + \lambda_4 \frac{d(\delta x_4)}{ds} + \lambda_5 \frac{d(\delta x_5)}{ds} \right) ds.$$

$$(A.27)$$

Integrating by parts the various terms of the last integral, we obtain

$$\int_0^1 \left(\boldsymbol{\lambda}_1^{\mathrm{T}} \frac{d(\delta \mathbf{x}_1)}{ds} \right) ds = \boldsymbol{\lambda}_1^{\mathrm{T}} \delta \mathbf{x}_1|_1 - \boldsymbol{\lambda}_1^{\mathrm{T}} \delta \mathbf{x}_1|_0 - \int_0^1 \left(\delta \mathbf{x}_1 \frac{d\boldsymbol{\lambda}_1^{\mathrm{T}}}{ds} \right) ds \qquad (A.28a)$$

$$\int_0^1 \left(\boldsymbol{\lambda}_2^{\mathrm{T}} \frac{d(\delta \mathbf{x}_2)}{ds} \right) ds = \boldsymbol{\lambda}_2^{\mathrm{T}} \delta \mathbf{x}_2|_1 - \boldsymbol{\lambda}_2^{\mathrm{T}} \delta \mathbf{x}_2|_0 - \int_0^1 \left(\delta \mathbf{x}_2 \frac{d\boldsymbol{\lambda}_2^{\mathrm{T}}}{ds} \right) ds \qquad (A.28b)$$

$$\int_0^1 \left(\lambda_3 \frac{d(\delta x_3)}{ds} \right) ds = \lambda_3 \delta x_3|_1 - \lambda_3 \delta x_3|_0 - \int_0^1 \left(\delta x_3 \frac{d\lambda_3}{ds} \right) ds \qquad (A.28c)$$

$$\int_0^1 \left(\lambda_4 \frac{d(\delta x_4)}{ds} \right) ds = \lambda_4 \delta x_4|_1 - \lambda_4 \delta x_4|_0 - \int_0^1 \left(\delta x_4 \frac{d\lambda_4}{ds} \right) ds \qquad (A.28d)$$

$$\int_0^1 \left(\lambda_5 \frac{d(\delta x_5)}{ds} \right) ds = \lambda_5 \delta x_5|_1 - \lambda_5 \delta x_5|_0 - \int_0^1 \left(\delta x_5 \frac{d\lambda_5}{ds} \right) ds. \qquad (A.28e)$$

Substituting the above equations (A.28a)–(A.28b) and using the initial conditions (A.9)–(A.12), the first and last terms of the rhs of equation (A.27) become

$$\frac{\partial G}{\partial \mathbf{x}_2} \delta \mathbf{x}_2|_1 - \int_0^1 \left(\boldsymbol{\lambda}_1^{\mathrm{T}} \frac{d(\delta \mathbf{x}_1)}{ds} + \boldsymbol{\lambda}_2^{\mathrm{T}} \frac{d(\delta \mathbf{x}_2)}{ds} + \lambda_3 \frac{d(\delta x_3)}{ds} + \lambda_4 \frac{d(\delta x_4)}{ds} + \lambda_5 \frac{d(\delta x_5)}{ds} \right) ds$$

$$= \frac{\partial G}{\partial \mathbf{x}_2} \delta \mathbf{x}_2|_1 - \boldsymbol{\lambda}_1^{\mathrm{T}} \delta \mathbf{x}_1|_1 + \boldsymbol{\lambda}_1^{\mathrm{T}} \delta \mathbf{x}_1|_0 - \boldsymbol{\lambda}_2^{\mathrm{T}} \delta \mathbf{x}_2|_1 + \boldsymbol{\lambda}_2^{\mathrm{T}} \delta \mathbf{x}_2|_0 - \lambda_4 \delta x_4|_1 - \lambda_5 \delta x_5|_1$$

$$+ \int_0^1 \left(\delta \mathbf{x}_1 \frac{d\boldsymbol{\lambda}_1^{\mathrm{T}}}{ds} + \delta \mathbf{x}_2 \frac{d\boldsymbol{\lambda}_2^{\mathrm{T}}}{ds} + \delta x_3 \frac{d\lambda_3}{ds} + \delta x_4 \frac{d\lambda_4}{ds} + \delta x_5 \frac{d\lambda_5}{ds} \right) ds. \qquad (A.29)$$

Now let us examine the terms

$$\frac{\partial G}{\partial \mathbf{x}_2} \delta \mathbf{x}_2|_1 - \boldsymbol{\lambda}_1^{\mathrm{T}} \delta \mathbf{x}_1|_1 + \boldsymbol{\lambda}_1^{\mathrm{T}} \delta \mathbf{x}_1|_0 - \boldsymbol{\lambda}_2^{\mathrm{T}} \delta \mathbf{x}_2|_1 + \boldsymbol{\lambda}_2^{\mathrm{T}} \delta \mathbf{x}_2|_0. \qquad (A.30)$$

From the initial conditions (A.7) and (A.8), we see that

$$\delta\mathbf{x}_2|_0 = \mathbf{0} \tag{A.31}$$

and

$$\delta\mathbf{x}_2|_1 = \mathbf{A}(\mathbf{J}_1\,\delta\mathbf{x}_2|_1 - \delta\mathbf{x}_1|_1) \tag{A.32}$$

where $\mathbf{J}_1$ is the Jacobian of $\mathbf{f}$ with respect to $\mathbf{x}_2$. Solving for $\delta\mathbf{x}_2|_1$, we get

$$(\mathbf{I} - \mathbf{A}\,\mathbf{J}_1)\,\delta\mathbf{x}_2|_1 = -\mathbf{A}\,\delta\mathbf{x}_1|_1$$

and then

$$\delta\mathbf{x}_2|_1 = -(\mathbf{I} - \mathbf{A}\,\mathbf{J}_1)^{-1}\mathbf{A}\,\delta\mathbf{x}_1|_1. \tag{A.33}$$

Substituting (A.31) and (A.33) in (A.30), we have

$$-\frac{\partial G}{\partial\mathbf{x}_2}(\mathbf{I} - \mathbf{A}\,\mathbf{J}_1)^{-1}\mathbf{A}\,\delta\mathbf{x}_1|_1 - \boldsymbol{\lambda}_1^{\mathrm{T}}\,\delta\mathbf{x}_1|_1 + \boldsymbol{\lambda}_1^{\mathrm{T}}\,\delta\mathbf{x}_1|_0 + \boldsymbol{\lambda}_2^{\mathrm{T}}(\mathbf{I} - \mathbf{A}\,\mathbf{J}_1)^{-1}\mathbf{A}\,\delta\mathbf{x}_1|_1. \tag{A.34}$$

We want this expression to be zero for all $\delta\mathbf{x}_1|_0$ and $\delta\mathbf{x}_1|_1$; thus we set

$$\boldsymbol{\lambda}_1^{\mathrm{T}} = 0 \qquad\qquad\qquad \text{at}\quad s = 0 \tag{A.35}$$

$$-\boldsymbol{\lambda}_1^{\mathrm{T}} - \frac{\partial G}{\partial\mathbf{x}_2}(\mathbf{I} - \mathbf{A}\,\mathbf{J}_1)^{-1}\mathbf{A} + \boldsymbol{\lambda}_2^{\mathrm{T}}(\mathbf{I} - \mathbf{A}\,\mathbf{J}_1)^{-1}\mathbf{A} = 0 \qquad \text{at}\quad s = 1. \tag{A.36}$$

Let us now elaborate (A.36):

$$\boldsymbol{\lambda}_2^{\mathrm{T}}(\mathbf{I} - \mathbf{A}\,\mathbf{J}_1)^{-1} = \boldsymbol{\lambda}_1^{\mathrm{T}}\mathbf{A}^{-1} + \frac{\partial G}{\partial\mathbf{x}_2}(\mathbf{I} - \mathbf{A}\,\mathbf{J}_1)^{-1}$$

$$\boldsymbol{\lambda}_2^{\mathrm{T}} = \boldsymbol{\lambda}_1^{\mathrm{T}}\mathbf{A}^{-1}(\mathbf{I} - \mathbf{A}\,\mathbf{J}_1) + \frac{\partial G}{\partial\mathbf{x}_2}$$

$$\boldsymbol{\lambda}_2^{\mathrm{T}} = \boldsymbol{\lambda}_1^{\mathrm{T}}\mathbf{A}^{-1} - \boldsymbol{\lambda}_1^{\mathrm{T}}\mathbf{J}_1 + \frac{\partial G}{\partial\mathbf{x}_2}$$

$$\boldsymbol{\lambda}_2 = (\mathbf{A}^{-1})^{\mathrm{T}}\boldsymbol{\lambda}_1 - \mathbf{J}_1^{\mathrm{T}}\boldsymbol{\lambda}_1 + \left(\frac{\partial G}{\partial\mathbf{x}_2}\right)^{\mathrm{T}}.$$

So the BCs for $\boldsymbol{\lambda}_1$ and $\boldsymbol{\lambda}_2$ are

$$\boldsymbol{\lambda}_1 = 0 \qquad\qquad\qquad \text{at}\quad s = 0 \tag{A.37}$$

$$\boldsymbol{\lambda}_2 = (\mathbf{A}^{-1})^{\mathrm{T}}\boldsymbol{\lambda}_1 - \mathbf{J}_1^{\mathrm{T}}\boldsymbol{\lambda}_1 + \left(\frac{\partial G}{\partial\mathbf{x}_2}\right)^{\mathrm{T}} \qquad \text{at}\quad s = 1. \tag{A.38}$$

Note that since for the *pellet* problem neither $\mathbf{f}$ nor G is a function of $\mathbf{x}_2$, the BC (A.38) becomes

$$\boldsymbol{\lambda}_2 = (\mathbf{A}^{-1})^{\mathrm{T}}\boldsymbol{\lambda}_1 \qquad \text{at}\quad s = 1$$

which is equivalent to

$$\lambda_2^T = \lambda_1^T \mathbf{A}^{-1}$$
$$\lambda_1^T = \lambda_2^T \mathbf{A} \tag{A.39}$$
$$\lambda_1 = \mathbf{A}^T \lambda_2$$

while for the *membrane* problem

$$\left(\frac{\partial G}{\partial \mathbf{x}_2}\right)^T = \mathbf{v}$$

where $\mathbf{v}$ is given by equations (B.42)–(B.44). Equation (A.27) now becomes

$$\delta I = \int_0^1 \left(\frac{\partial H}{\partial \mathbf{x}_1} + \frac{d\lambda_1^T}{ds}\right) \delta \mathbf{x}_1 \, ds + \int_0^1 \left(\frac{\partial H}{\partial \mathbf{x}_2} + \frac{d\lambda_2^T}{ds}\right) \delta \mathbf{x}_2 \, ds$$

$$+ \int_0^1 \left(\frac{\partial H}{\partial x_3} + \frac{d\lambda_3}{ds}\right) \delta x_3 \, ds + \int_0^1 \left(\frac{\partial H}{\partial x_4} + \frac{d\lambda_4}{ds}\right) \delta x_4 \, ds$$

$$+ \int_0^1 \left(\frac{\partial H}{\partial x_5} + \frac{d\lambda_5}{ds}\right) \delta x_5 \, ds + \int_0^1 \frac{\partial H}{\partial \mu} \delta \mu \, ds$$

$$+ \left(\frac{\partial G}{\partial x_4} - \lambda_4\right) \delta x_4|_1 + \left(\frac{\partial G}{\partial x_5} - \lambda_5\right) \delta x_5|_1 \tag{A.40}$$

where

$$H = R + \lambda_1^T \mathbf{f}_1 + \lambda_2^T \mathbf{f}_2 + \lambda_3 f_3 + \lambda_4 f_4 + \lambda_5 f_5 \tag{A.41}$$

is called the *Hamiltonian*. In order to express the direct influence of $\delta \mu$ alone, let us choose the arbitrary functions λ such that they satisfy

$$\frac{d\lambda_1^T}{ds} = -\frac{\partial H}{\partial \mathbf{x}_1} \tag{A.42}$$

$$\frac{d\lambda_2^T}{ds} = -\frac{\partial H}{\partial \mathbf{x}_2} \tag{A.43}$$

$$\frac{d\lambda_3}{ds} = -\frac{\partial H}{\partial x_3} \tag{A.44}$$

$$\frac{d\lambda_4}{ds} = -\frac{\partial H}{\partial x_4} \tag{A.45}$$

$$\frac{d\lambda_5}{ds} = -\frac{\partial H}{\partial x_5}. \tag{A.46}$$

This allows the influence of the system equations (A.1)–(A.5) to be transmitted by

λ and to be felt in $\partial H/\partial\mu$. For the terms outside the integral in (A.40) we choose

$$\lambda_4 = \frac{\partial G}{\partial x_4} \quad \text{at} \quad s = 1 \tag{A.47}$$

$$\lambda_5 = \frac{\partial G}{\partial x_5} \quad \text{at} \quad s = 1. \tag{A.48}$$

Thus, on using equations (A.42)–(A.48), equation (A.40) becomes

$$\delta I = \int_0^1 \frac{\partial H}{\partial\mu} \delta\mu \, ds. \tag{A.49}$$

Equation (A.49) represents the direct influence of the variation $\delta\mu$ on δI. The necessary condition for optimality is that $\delta I \leq 0$ for all possible small variations $\delta\mu$, which implies that

$$\frac{\partial H}{\partial\mu} = 0. \tag{A.50}$$

This condition is used in sections 2.5.1 and 5.2 to identify the optimal catalyst distribution $\mu(s)$ in various cases.

APPENDIX B

Optimal Catalyst Distribution in Pellets for an Inert Membrane Reactor: Problem Formulation

In this appendix, equations describing the mass and energy balances in pellets and in both feed and permeate-side fluid streams of an inert membrane reactor, shown in Figure 5.1, are presented (cf. Yeung et al., 1994). The problem of determining the optimal catalyst distribution in pellets is then formulated. Optimization of the catalyst loading profile is carried out, as in section 2.5.1, using as optimization variable the dimensionless loading of the active catalyst

$$\mu(s) = \frac{q(s)}{\bar{q}} \tag{B.1}$$

which, by definition, satisfies the condition

$$\int_0^1 \mu(s) s^n \, ds = \frac{1}{n+1}. \tag{B.2}$$

On physical grounds, it is assumed that $\mu(s)$ is positive and bounded, i.e.

$$0 \le \mu(s) < \alpha \tag{B.3}$$

where α represents the maximum allowable value of the local catalyst loading on the support.

The dependence of the reaction rate on catalyst loading is represented by equation (2.129):

$$r = r_0 \frac{\mu(s)}{1 + B\mu(s)}. \tag{B.4}$$

B.1   The Mass and Energy Balance Equations

The dimensionless steady-state mass and energy balances for the *pellet*, taking account of external transport resistances, are given by

$$L[u_i] = \frac{1}{s^n} \frac{d}{ds}\left(s^n \frac{du_i}{ds}\right) = \frac{\mu(s)}{1 + B\mu(s)} D_i \sum_{j=1}^{J} v_{ij} \phi_j^2 f_j(\mathbf{u}, \theta) \qquad (i = 1, \ldots, I) \tag{B.5}$$

$$L[\theta] = \frac{1}{s^n}\frac{d}{ds}\left(s^n\frac{d\theta}{ds}\right) = -\frac{\mu(s)}{1+B\mu(s)}\sum_{j=1}^{J}\beta_j\phi_j^2 f_j(\mathbf{u},\theta) \qquad (B.6)$$

with BCs

$$s = 0: \qquad \frac{du_i}{ds} = 0 \qquad\qquad (i = 1,\ldots, I) \qquad (B.7a)$$

$$s = 1: \qquad \frac{du_i}{ds} = \mathrm{Bi}_{m,i}(g_i - u_i) \qquad (i = 1,\ldots, I) \qquad (B.7b)$$

$$s = 0: \qquad \frac{d\theta}{ds} = 0 \qquad\qquad\qquad (B.7c)$$

$$s = 1: \qquad \frac{d\theta}{ds} = \mathrm{Bi}_h(\tau - \theta) \qquad\qquad (B.7d)$$

where the following dimensionless parameters are used:

$$u_i = C_i/C_{f,1}^0, \qquad s = x/R, \qquad D_i = D_{e,1}/D_{e,i}$$

$$g_i = C_{f,i}/C_{f,1}^0, \qquad \tau = T_f/T_f^0$$

$$\phi_j^2 = r_j\left(C_{f,1}^0, C_{f,1}^0, \quad\ldots, C_{f,1}^0, T_f^0\right) R^2/D_{e,1}C_{f,1}^0$$

$$f_j(\mathbf{u},\theta) = r_j(C_1, C_2, \ldots, C_I, T)/r_j\left(C_{f,1}^0, C_{f,1}^0, \ldots, C_{f,1}^0, T_f^0\right)$$

$$\theta = T/T_f^0, \qquad\qquad \beta_j = (-\Delta H_j)D_{e,1}C_{f,1}^0/(\lambda_e T_f^0)$$

$$\mathrm{Bi}_{m,i} = k_{g,i}R/D_{e,i}, \quad \mathrm{Bi}_h = hR/\lambda_e. \qquad\qquad (B.8)$$

The mass and energy balances can be rewritten in matrix form as follows:

$$L[\mathbf{x}] = \frac{\mu(s)}{1+B\mu(s)}\,\mathbf{w}(\mathbf{x}) \qquad (B.9)$$

with BCs

$$s = 0: \qquad \frac{d\mathbf{x}}{ds} = \mathbf{0} \qquad\qquad (B.10a)$$

$$s = 1: \qquad \frac{d\mathbf{x}}{ds} = \mathbf{A}(\mathbf{x}_f - \mathbf{x}) \qquad (B.10b)$$

where the vectors $\mathbf{x}$, $\mathbf{x}_f$, and $\mathbf{w}$, and the matrix $\mathbf{A}$, are defined by

$$\mathbf{x} = \begin{bmatrix} u_1 \\ u_2 \\ \vdots \\ u_I \\ \theta \end{bmatrix}, \quad \mathbf{w}(\mathbf{x}) = \begin{bmatrix} D_1\sum_{j=1}^{J}\nu_{1j}\phi_j^2 f_j(\mathbf{x}) \\ D_2\sum_{j=1}^{J}\nu_{2j}\phi_j^2 f_j(\mathbf{x}) \\ \vdots \\ D_I\sum_{j=1}^{J}\nu_{Ij}\phi_j^2 f_j(\mathbf{x}) \\ -\sum_{j=1}^{J}\beta_j\phi_j^2 f_j(\mathbf{x}) \end{bmatrix}, \quad \mathbf{x}_f = \begin{bmatrix} g_1 \\ g_2 \\ \vdots \\ g_I \\ \tau \end{bmatrix} \qquad (B.11)$$

$$
\mathbf{A} = \begin{bmatrix}
\text{Bi}_{m,1} & 0 & & & & \\
0 & \text{Bi}_{m,2} & 0 & & & \\
 & \ddots & \ddots & \ddots & & \\
 & & 0 & \text{Bi}_{m,I} & 0 & \\
 & & & 0 & \text{Bi}_h
\end{bmatrix}.
\tag{B.12}
$$

The steady-state mass balances for the cylindrical *inert membrane*, accounting for external transport resistances, are given by

$$
D_{\text{eM},i} \frac{1}{x_M} \frac{d}{dx_M} \left(x_M \frac{dC_{M,i}}{dx_M} \right) = 0 \qquad (i = 1, \ldots, I)
\tag{B.13}
$$

with BCs

$$
x_M = x_F: \qquad -D_{\text{eM},i} \frac{dC_{M,i}}{dx_M} = k_{g,F}(C_{F,i} - C_{M,i})
\tag{B.14a}
$$

$$
x_M = x_P: \qquad -D_{\text{eM},i} \frac{dC_{M,i}}{dx_M} = k_{g,P}(C_{M,i} - C_{P,i}).
\tag{B.14b}
$$

The mass flux of component i, at the inner or outer membrane surface, can be obtained on solving the above equations:

$$
-2\pi x_P L D_{\text{eM},i} \left(\frac{dC_{M,i}}{dx_M} \right)_{x_M = x_P} = -2\pi x_F L D_{\text{eM},i} \left(\frac{dC_{M,i}}{dx_M} \right)_{x_M = x_F}
$$

$$
= \frac{2\pi L D_{\text{eM},i} \left(\frac{\zeta_{F,i} P_F}{R_g T_F} - \frac{\zeta_{P,i} P_P}{R_g T_P} \right)}{\ln \frac{x_P}{x_F} + D_{\text{eM},i} \left(\frac{1}{k_{g,F} x_F} + \frac{1}{k_{g,P} x_P} \right)}.
\tag{B.15}
$$

Similarly, for the heat flow across the membrane,

$$
-2\pi x_P L \lambda_{\text{eM}} \left(\frac{dT_M}{dx_M} \right)_{x_M = x_P} = -2\pi x_F L \lambda_{\text{eM}} \left(\frac{dT_M}{dx_M} \right)_{x_M = x_F}
$$

$$
= \frac{2\pi L \lambda_{\text{eM}} (T_F - T_P)}{\ln \frac{x_P}{x_F} + \lambda_{\text{eM}} \left(\frac{1}{h_F x_F} + \frac{1}{h_P x_P} \right)}.
\tag{B.16}
$$

The mass balances for the *feed side* of the membrane reactor are given by

$$
Q_F^0 \zeta_{F,i}^0 - Q_F \zeta_{F,i} - \frac{V_r (1 - \varepsilon)(n + 1) D_{e,i} C_{F,1}^0}{R^2} \frac{du_i(1)}{ds}
$$

$$
- \frac{2\pi L D_{\text{eM},i} \left(\frac{\zeta_{F,i} P_F}{R_g T_F} - \frac{\zeta_{P,i} P_P}{R_g T_P} \right)}{\ln \frac{x_P}{x_F} + D_{\text{eM},i} \left(\frac{1}{k_{g,F} x_F} + \frac{1}{k_{g,P} x_P} \right)} = 0
\tag{B.17}
$$

and for the *permeate side* by

$$
Q_P^0 \zeta_{P,i}^0 - Q_P \zeta_{P,i} + \frac{2\pi L D_{\text{eM},i} \left(\frac{\zeta_{F,i} P_F}{R_g T_F} - \frac{\zeta_{P,i} P_P}{R_g T_P} \right)}{\ln \frac{x_P}{x_F} + D_{\text{eM},i} \left(\frac{1}{k_{g,F} x_F} + \frac{1}{k_{g,P} x_P} \right)} = 0.
\tag{B.18}
$$

Additionally, we have the two congruence equations arising from the definition of the mole fractions in the feed and permeate streams,

$$\zeta_{F,i} = \frac{C_{F,i}}{\sum_{i=1}^{I} C_{F,i}} \qquad \zeta_{P,i} = \frac{C_{P,i}}{\sum_{i=1}^{I} C_{P,i}} \tag{B.19}$$

given by

$$\sum_{i=1}^{I} \zeta_{F,i} = 1 \quad \text{and} \quad \sum_{i=1}^{I} \zeta_{P,i} = 1. \tag{B.20}$$

The energy balance for the *feed side* of the membrane reactor is

$$\left(\rho c_p Q_F^0\right)_{T_F^0} T_F^0 - \left(\rho c_p Q_F\right)_{T_F} T_F - \frac{V_r(1-\varepsilon)(n+1)\lambda_e T_F^0}{R^2} \frac{d\theta(1)}{ds}$$

$$- \frac{2\pi L \lambda_{eM}(T_F - T_P)}{\ln \frac{x_P}{x_F} + \lambda_{eM}\left(\frac{1}{h_F x_F} + \frac{1}{h_P x_P}\right)} = 0 \tag{B.21}$$

and for the *permeate side*

$$\left(\rho c_p Q_P^0\right)_{T_P^0} T_P^0 - \left(\rho c_p Q_P\right)_{T_P} T_P - \frac{2\pi L \lambda_{eM}(T_F - T_P)}{\ln \frac{x_P}{x_F} + \lambda_{eM}\left(\frac{1}{h_F x_F} + \frac{1}{h_P x_P}\right)} = 0. \tag{B.22}$$

Equations (B.17)–(B.22) represent $2I + 4$ linearly independent equations in the $2I + 4$ unknowns, $\zeta_{F,1}, \zeta_{F,2}, \ldots, \zeta_{F,I}, \zeta_{P,1}, \zeta_{P,2}, \ldots, \zeta_{P,I}, Q_F, Q_P, T_F,$ and T_P. By careful examination, we can see that these equations can be solved once the values of the derivatives $du_i(1)/ds$ and $d\theta(1)/ds$ are known. This means that once the fluxes leaving the catalyst pellets are known, equations (B.17)–(B.22) provide the values for composition and temperature at the outlet of both feed and permeate sides. We can formally state this conclusion as follows:

$$\zeta_F = \mathbf{p}_1[\gamma(1)] \tag{B.23}$$

$$\zeta_P = \mathbf{p}_2[\gamma(1)] \tag{B.24}$$

$$\mathbf{x}_f = \mathbf{p}_3[\gamma(1)] \tag{B.25}$$

where $\mathbf{p}$ indicates a functional dependence and $\gamma(1) = d\mathbf{x}(1)/ds$ with $\mathbf{x}$ defined by equation (B.11).

B.2   The Performance Indexes

The performance indexes of interest for optimization are:
Conversion of reactant species N:

$$X_N = \frac{Q_F^0 \zeta_{F,N}^0 + Q_P^0 \zeta_{P,N}^0 - Q_F \zeta_{F,N} - Q_P \zeta_{P,N}}{Q_F^0 \zeta_{F,N}^0 + Q_P^0 \zeta_{P,N}^0} \tag{B.26}$$

which, using equations (B.17) and (B.18), becomes

$$X_N = \frac{K_N \gamma_N(1)}{Q_F^0 \zeta_{F,N}^0 + Q_P^0 \zeta_{P,N}^0} \tag{B.27}$$

where

$$K_N = -\frac{V_r(1-\varepsilon)(n+1)D_{e,N}C_{F,1}^0}{R^2}. \tag{B.28}$$

Selectivity of a product D relative to a reactant N:

$$S_{DN} = \frac{Q_F \zeta_{F,D} + Q_P \zeta_{P,D}}{Q_F^0 \zeta_{F,N}^0 + Q_P^0 \zeta_{P,N}^0 - Q_F \zeta_{F,N} - Q_P \zeta_{P,N}} \tag{B.29}$$

which, using a substitution similar to that for conversion, leads to

$$S_{DN} = \frac{Q_F^0 \zeta_{F,D}^0 + Q_P^0 \zeta_{P,D}^0 - K_D \gamma_D(1)}{K_N \gamma_N(1)}. \tag{B.30}$$

Purity of a product D relative to a reactant N in the feed side:

$$P_{DN} = \frac{\zeta_{F,D}}{\zeta_{F,N}} \tag{B.31}$$

which, using equation (B.23), reduces to

$$P_{DN} = \frac{p_{1D}[\gamma(1)]}{p_{1N}[\gamma(1)]} = D[\gamma(1)]. \tag{B.32}$$

B.3   Development of the Hamiltonian

Equations (B.2), (B.9), (B.10) for the pellet are equivalent to the following set of first-order differential equations:

$$\frac{d\mathbf{x}}{ds} = \gamma s^{-n} \tag{B.33}$$

$$\frac{d\gamma}{ds} = \frac{\mu(s)}{1 + B\mu(s)} \mathbf{w}(\mathbf{x}) s^n \tag{B.34}$$

$$\frac{dh}{ds} = \mu(s) s^n \tag{B.35}$$

with BCs

$$s = 0: \qquad \gamma = \mathbf{0} \tag{B.36a}$$

$$s = 1: \qquad \gamma = \mathbf{A}(\mathbf{x_f} - \mathbf{x}) \tag{B.36b}$$

$$s = 0: \qquad h = 0 \tag{B.36c}$$

$$s = 1: \qquad h = \frac{1}{n+1}. \tag{B.36d}$$

We can now introduce the Hamiltonian, whose expression for all performance indexes is

$$II = \frac{\mu(s)}{1 + B\mu(s)}\left[\lambda_2^T \mathbf{w}(\mathbf{x})\right]s^n + \lambda_1^T \gamma s^{-n} + \lambda_3\, \mu(s)\, s^n \tag{B.37}$$

where λ_1 and λ_2 (column vectors, with $I+1$ elements) and λ_3 are the Lagrange multipliers defined by the following differential equations:

$$\frac{d\lambda_1}{ds} = -\frac{\mu(s)}{1 + B\mu(s)}\, s^n [\mathbf{J}^T \lambda_2] \tag{B.38}$$

$$\frac{d\lambda_2}{ds} = -\lambda_1\, s^{-n} \tag{B.39}$$

$$\frac{d\lambda_3}{ds} = 0 \tag{B.40}$$

where $\mathbf{J}$ is the Jacobian matrix of the vector $\mathbf{w}$ with respect to the independent variable $\mathbf{x}$. The relevant BCs for the Lagrange multipliers are

$$s = 0: \qquad \lambda_1 = \mathbf{0} \tag{B.41a}$$

$$s = 1: \qquad \lambda_2 = (\mathbf{A}^{-1})^T \lambda_1 - \mathbf{J}_1^T \lambda_1 + \mathbf{v} \tag{B.41b}$$

where $\mathbf{J}_1$ is the Jacobian matrix of $\mathbf{p}_3$ with respect to γ, and $\mathbf{v}$ is a column vector of $I + 1$ elements which depends on the objective functional being maximized as follows:

For conversion maximization

$$v_i = 0 \qquad \text{for}\quad i \neq N \tag{B.42a}$$

$$v_N = \frac{K_N}{Q_F^0 \zeta_{F,N}^0 + Q_P^0 \zeta_{P,N}^0}. \tag{B.42b}$$

For selectivity maximization

$$v_i = 0 \qquad \text{for}\quad i \neq N,\, D \tag{B.43a}$$

$$v_N = -\frac{Q_F^0 \zeta_{F,D}^0 + Q_P^0 \zeta_{P,D}^0 - K_D \gamma_D(1)}{K_N \gamma_N^2(1)} \tag{B.43b}$$

$$v_D = -\frac{K_D}{K_N \gamma_N(1)}. \tag{B.43c}$$

For product purity maximization

$$\mathbf{v} = \mathbf{d}_\gamma \tag{B.44}$$

where $\mathbf{d}_\gamma$ is the column vector of the partial derivatives of D with respect to γ.

Having formulated the problem, the necessary condition for optimality,

$$\frac{\partial H}{\partial \mu} = 0 \tag{B.45}$$

is derived in Appendix A.

Notation

A	surface area of active element per unit catalyst pellet weight (m^2 per gram of catalyst pellet)
$\mathbf{A}$	matrix defined by equation (2.102)
A_i	component involved in a reaction
a	dimensionless catalyst activity
a_p	pore surface area per unit volume
B	$b\bar{q}$
Bi^*	dimensionless parameter defined by equation (2.32)
Bi_h	Biot number for heat transfer, hR/λ_e
Bi_m	Biot number for mass transfer, $k_g R/D_e$
b	constant defined by equation (2.122)
b_1, b_2	constants defined by equation (3.27)
C	concentration
$\mathbf{c}$	concentration vector
c	capacitance
c_p	fluid heat capacity
D	catalyst dispersion
D_e	effective diffusivity
D_i	dimensionless effective diffusivity
D_0	catalyst dispersion as loading $q \to 0$
Da	Damköhler number
d	number of zones in a reactor
E	activation energy
e	reactant consumption rate defined by equation (3.11)
e	elementary charge
F	Faraday constant or flux
f, f_j	dimensionless reaction rate
$\bar{f}$	dimensionless production rate of desired product, integrated over the catalyst pellet
G^*	function defined by equation (2.80)
$\mathcal{G}$	performance index for deactivating catalyst defined by equation (4.12)

g	dimensionless concentration in the external fluid phase
g_v	volume-averaged catalyst density
H	Hamiltonian functional
$\mathcal{H}$	average reactant conversion over catalyst lifetime defined by equation (4.5)
h	heat transfer coefficient or function defined by equations (2.137), (2.160), (2.187)
I	ionic strength
I_0, I_1	Bessel functions
$\mathbf{J}$	Jacobian matrix defined by equation (2.144)
J_c	capillary flow rate
J_d	diffusive transport rate
J_v	evaporation rate
K	equilibrium constant
$\mathcal{K}$	performance index for deactivating catalyst defined by equation (4.13)
K_g	permeability of gel layer
K_m	Michaelis–Menten constant
K^{sol}	equilibrium adsorption constant due to solution effects
K^{st}	equilibrium adsorption constant due to steric hinderance
k	local reaction rate constant
$\bar{k}$	volume-averaged reaction rate constant
k^+	adsorption rate constant
k^-	desorption rate constant
k_B	Boltzmann constant
k_g	mass transfer coefficient
k^{sol}	desorption rate constant due to solution effects
L	reactor length
$L[\ ]$	differential operator: $(\frac{1}{s^n})\{\frac{d}{ds}(s^n\frac{d[\ ]}{ds})\}$
L_g	gel layer thickness
M	maximum value of dimensionless reaction rate defined by equation (2.16)
m	reaction order or constant defined by equation (7.3)
N_A	Avogadro's number
N_p	number of pellets
N_s	surface site density
N_X	surface area of oxide per unit solution volume
n	integer, characteristic of pellet geometry: $= 0$ for infinite slab, $= 1$ for infinite cylinder, $= 2$ for sphere; or surface concentration
P	pressure
p	specific surface area of active element (m^2 per gram of active element)
Q	molar flow rate
q	local weight fraction of catalyst (grams of active element per gram of catalyst pellet)
$\bar{q}$	volume-averaged weight fraction of catalyst
R	characteristic dimension: half thickness ($n = 0$), radius ($n = 1, 2$)
R_g	universal gas constant
R_m	reactant molecule radius
R_p	pore radius

r, r_j, r'	reaction rate
$\bar{r}$	volume-averaged reaction rate defined by equation (3.3)
r_f	position of imbibition front
r_0	reaction rate in a catalyst pellet where the active surface area per unit catalyst weight is equal to $p\bar{q}$
S	selectivity
St	Stanton number
s	dimensionless radial coordinate
$\bar{s}$	dimensionless location of catalyst in pellet
T	temperature
t	time
U_1, U_2	objective functions defined by equations (2.108), (2.109)
u	dimensionless solid-phase concentration
$\bar{u}$	dimensionless concentration at the catalyst location
$\mathbf{u}$	dimensionless solid-phase concentration vector
V	volume
V_p	pellet volume
v	velocity or bond valence
W	catalyst weight or mass flux
w	dimensionless feed distribution along a reactor
$\mathbf{w}$	vector defined by equation (2.101)
$\mathbf{w}_{Mx}$	vector defined by equation (2.143)
X	reactant conversion
X_T	total conversion for membrane reactor defined by equation (5.1)
x	radial coordinate
$\mathbf{x}$	vector defined by equation (2.101)
$\mathbf{x}_f$	vector defined by equation (2.101)
Y	yield
y	dimensionless axial coordinate
$\hat{y}$	reactor portion with subsurface optimal location or reactor portion with distributed feed
Z	ion valence
z	axial coordinate
$[\ ]$	concentration

Greek Letters

α	upper bound of the dimensionless active-element concentration defined by equation (2.128)
α_1	weighting coefficient proportional to value of products
α_2	weighting coefficient proportional to cost of raw materials
α_3	weighting coefficient proportional to cost of catalyst
α_g	constant [see equation (5.14)]
β	dimensionless heat of reaction or fraction of sulfonic groups
$\bar{\beta}$	external Prater number

Γ	dimensionless molar flow rate
γ	dimensionless activation energy or perturbation [see equation (2.84)]
$\boldsymbol{\gamma}$	vector defined by equation (2.135)
Δ	dimensionless half thickness of step distribution
ΔH	heat of reaction
ΔH_a	heat of adsorption
ΔP_g	pressure drop in gel layer
ΔT	temperature rise
δ	deposition thickness
δ_M	membrane thickness
$\delta(s - \bar{s})$	Dirac-delta function at $\bar{s}$
δa	perturbation of catalyst distribution
δu	perturbation of dimensionless concentration
$\delta \eta$	perturbation of effectiveness factor
ε	dimensionless heat of adsorption or void fraction
ε_0	permittivity of vacuum
ε_r	relative permittivity of solvent
ζ	mole fraction
η	effectiveness factor or viscosity
Θ	dimensionless time: for membrane reactor defined by equation (5.3)
θ	dimensionless temperature
$\bar{\theta}$	dimensionless temperature at the catalyst location
κ	ratio of Thiele moduli, ϕ_2^2/ϕ_1^2
Λ	dimensionless parameter defined by equation (2.26)
Λ_e	dimensionless parameter defined by equation (2.31)
Λ'	dimensionless parameter defined by equation (2.48)
Λ'_e	dimensionless parameter defined by equation (2.56)
λ	Lagrange multiplier
λ_e	effective thermal conductivity
μ	dimensionless active element concentration, or Bi_m/Bi_h
ν_{ij}	stoichiometric coefficient of the ith component in the jth reaction
ξ	dimensionless coordinate in a membrane
$\bar{\xi}$	dimensionless catalyst location in a membrane
Π	dimensionless parameter defined by equation (3.14)
π	empirical coefficient defined by equation (4.4)
ρ	density
ρ_d	deposited layer density
ρ_s	pellet density
σ	dimensionless adsorption equilibrium constant, or electric charge density
σ_c	critical value of dimensionless adsorption equilibrium constant defined by equation (2.47)
τ	dimensionless temperature in the external fluid phase
υ	volumetric flow rate
Φ	switching function
ϕ	Thiele modulus

ϕ_0	"clean" Thiele modulus defined by equation (2.24)
Ψ	electrostatic potential
ψ	modified Thiele modulus, $\phi/\sqrt{B}$
ψ_n	function of catalyst location defined by equation (2.121)
Ω	function defined by equation (2.112)
ω	function defined by equation (2.173)

Subscripts

0	0 plane; initial
b	bulk liquid
c	coolant
d	diffuse layer
e	equilibrium
ext	external
F	feed section of a membrane reactor
f	fluid phase
IHP	inner Helmholz plane
i	component i
ip	for product inhibition
is	for substrate inhibition
j	reaction j
M	membrane
m, max	maximum
OHP	outer Helmholz plane
oper	operating
opt	optimal
P	permeate
p	poison or pore
r	reactor
s	saturation or surface
un	uniform

Superscripts

| 0 | reactor inlet |
| * | value corresponding to the optimal catalyst distribution |

References

Abello, M. C., Velasco, A. P., Gorriz, O. F., and Rivarola, J. B. 1995. Temperature-programmed desorption study of the acidic properties of γ-alumina, *Appl. Catal. A.* 129: 93–100.

Aben, P. C. 1968. Palladium areas in supported catalysts. Determination of palladium surface areas in supported catalysts by means of hydrogen chemisorption, *J. Catal.* 10: 224–229.

Adcock, D. S., and McDowall, I. C. 1957. The mechanism of filter pressing and slip casting, *J. Amer. Ceram. Soc.* 40: 355–362.

Agashe, K. B., and Regalbuto, J. R. 1997. A revised physical theory for adsorption of metal complexes at oxide surfaces, *J. Coll. Interf. Sci.* 185: 174–189.

Ahn, J. H., Ihm, S. K., and Park, K. S. 1988. The effects of the local concentration and distribution of sulfonic acid groups on 1-butene isomerization catalyzed by macroporous ion-exchange resin catalyst, *J. Catal.* 113: 434–443.

Akratopulu, K. C., Vordonis, L., and Lycourghiotis, A. 1986. Effect of temperature on the point of zero charge and surface dissociation constants in aqueous suspensions of γ-Al_2O_3, *J. Chem. Soc., Faraday, Trans. I* 82: 3697–3708.

Akratopulu, K. C., Vordonis, L., and Lycourghiotis, A. 1988. Development of carriers with controlled concentration of charged surface groups in aqueous solutions. III. Regulation of the point of zero charge, surface dissociation constants and concentration of charged surface groups of SiO_2 by variation of the solution temperature or by modification with sodium ions, *J. Catal.* 109: 41–50.

Alexiou, M. S., and Sermon, P. A. 1993. Aspects of the preparation of heterogeneous catalysts by impregnation, *React. Kinet. Catal. Lett.* 51: 1–7.

Anderson, J. R. 1975. *Structure of Metallic Catalysts*, London: Academic Press.

Arakawa, H., Takeuchi, K., Matsuzaki, T., and Sugi, Y. 1984. Effect of metal dispersion on the activity and selectivity of Rh/SiO_2 catalyst for high pressure CO hydrogenation, *Chem. Lett.* 1607–1610.

Arbabi, S., and Sahimi, M. 1991a. Computer simulations of catalyst deactivation – I. Model formulation and validation, *Chem. Eng. Sci.* 46: 1739–1747.

Arbabi, S., and Sahimi, M. 1991b. Computer simulations of catalyst deactivation – II. The effect of morphological transport and kinetic parameters on the performance of the catalyst, *Chem. Eng. Sci.* 46: 1749–1755.

Ardiles, D. R. 1986. Activity and selectivity of nonuniform bifunctional catalysts. Analysis of the fixed-bed reactor performance, *Collect. Czech. Chem. Commun.* 51: 2509–2520.

Ardiles, D. R., Scelza, O. A., and Castro, A. A. 1985. Activity and selectivity of non-uniform bifunctional catalysts, *Collect. Czech. Chem. Commun.* 50: 726–737.

Ardiles, D. R., De Miguel, S. R., Castro, A. A., and Scelza, O. A. 1986. Pt–Re catalysts: Study of the impregnation step, *Appl. Catal.* 24: 175–186.

Aris, R. 1975. *The Mathematical Theory of Diffusion and Reaction in Permeable Catalysts*, Oxford: Clarendon Press.

Asua, J. M., and Delmon, B. 1987. Theoretical study of the influence of nonuniform active-phase distribution on activity and selectivity of hydrodesulfurization catalysts, *Ind. Eng. Chem. Res.* 26: 32–39.

Au, S. S., Dranoff, J. S., and Butt, J. B. 1995. Nonuniform activity distribution in catalyst particles: Benzene hydrogenation on supported nickel in a single pellet diffusion reactor, *Chem. Eng. Sci.* 50: 3801–3812.

Bacaros, T., Bebelis, S., Pavlou, S., and Vayenas, C. G. 1987. Optimal catalyst distribution in pellets with shell progressive poisoning: The case of linear kinetics. In *Catalyst Deactivation 1987*, eds. Delmon, B., and Froment, G. F., pp. 459–468. Amsterdam: Elsevier.

Baiker, A., and Holstein, W. L. 1983. Impregnation of alumina with copper chloride – modeling of impregnation kinetics and internal copper profiles, *J. Catal.* 84: 178–188.

Baiker, A., Monti, D., and Wokaun, A. 1986. Impregnation of alumina with copper chloride: Evidence for differently immobilized copper species, *Appl. Catal.* 23: 425–436.

Baratti, R., Cao, G., Morbidelli, M., and Varma, A. 1990. Optimal activity distribution in nonuniformly impregnated catalyst particles: Numerical analysis, *Chem. Eng. Sci.* 45: 1643–1646.

Baratti, R., Wu, H., Morbidelli, M., and Varma, A. 1993. Optimal catalyst activity profiles in pellets – X. The role of catalyst loading, *Chem. Eng. Sci.* 48: 1869–1881.

Baratti, R., Gavriilidis, A., Morbidelli, M., and Varma, A. 1994. Optimization of a non-isothermal nonadiabatic fixed-bed reactor using Dirac-type silver catalysts for ethylene epoxidation, *Chem. Eng. Sci.* 49: 1925–1936.

Baratti, R., Feckova, V., Morbidelli, M., and Varma, A. 1997. Optimal catalyst activity profiles in pellets. 11. The case of multiple-step distributions, *Ind. Eng. Chem. Res.* 36: 3416–3420.

Barbier, J., and Marecot, P. 1981. Comparative study of structure sensitive reactions on Pt/Al_2O_3 and Ir/Al_2O_3 catalysts, *Nouveau J. Chim.* 5: 393–396.

Barrow, N. J., and Bowden, J. W. 1987. A comparison of models for describing the adsorption of anions on a variable charge mineral surface, *J. Coll. Interf. Sci.* 119: 236–250.

Bartholomew, C. H. 1993. Sintering kinetics of supported metals: New perspectives from a unifying GPLE treatment, *Appl. Catal. A* 107: 1–57.

Bartholomew, C. H., and Farrauto, R. J. 1976. Chemistry of nickel–alumina catalysts, *J. Catal.* 45: 41–53.

Bartholomew, C. H., Pannell, R. B., and Butler, J. L. 1980. Support and crystallite size effects in CO hydrogenation on nickel, *J. Catal.* 65: 335–347.

Barto, M., Markos, J., and Brunovska, A. 1991. Oscillatory behaviour of a catalyst pellet with narrow region of activity, *Chem. Eng. Sci.* 46: 2875–2880.

Basset, J. M., Dalmai-Imelik, G., Primet, M., and Mutin, R. 1975. A study of benzene hydrogenation and identification of the adsorbed species with Pt/Al_2O_3 catalysts, *J. Catal.* 37: 22–36.

Beck, D. D., Sommers, J. W., and DiMaggio C. L. 1997. Axial characterization of catalytic activity in close-coupled lightoff and underfloor catalytic converters, *Appl. Catal. B* 11: 257–272.

Becker, E. R., and Nuttall, T. A. 1979. Controlled catalyst distribution on supports by co-impregnation. In *Preparation of Catalysts II*, eds. Delmon, B., Grange, P., Jacobs, P., and Poncelet, G., pp. 159–169. Amsterdam: Elsevier.

Becker, E. R., and Wei, J. 1977a. Nonuniform distribution of catalysts on supports. I. Bimolecular Langmuir reactions, *J. Catal.* 46: 365–371.

Becker, E. R., and Wei, J. 1977b. Nonuniform distribution of catalysts on supports. II. First order reactions with poisoning, *J. Catal.* 46: 372–381.

Beeckman, J. W., and Hegedus, L. L. 1991. Design of monolith catalysts for power plant NO_x emission control, *Ind. Eng. Chem. Res.* 30: 969–978.

Benesi, H. A., Curtis, R. M., and Studer, H. P. 1968. Preparation of highly dispersed catalytic metals. Platinum supported on silica gel, *J. Catal.* 10: 328–335.

Benjamin, M. M., and Leckie, J. O. 1981. Multiple-site adsorption of Cd, Cu, Zn and Pb on amorphous iron oxyhydroxide, *J. Coll. Interf. Sci.* 79: 209–221.

Berenblyum, A. S., Mund, S. L., Karelskii, V. V., Goranskaya, T. P., Zolotukhin, V. D., and Lakhman, L. I. 1986. Catalysts for the selective hydrogenation of unsaturated compounds consisting of alumina particles with a regular palladium distribution, *Kinet. and Catal.* 27: 184–187.

Berger, D., and Pei, D. C. T. 1973. Drying of hygroscopic capillary porous solids – a theoretical approach, *Int. J. Heat Mass Transfer* 16: 293–302.

Blachou, V. M., and Philippopoulos, C. J. 1993. Adsorption of hexachloroplatinic acid on γ-alumina coatings for preparation of monolithic structure catalysts, *Chem. Eng. Commun.* 119: 41–53.

Blanco, M. N., Caceres, C. V., Fierro, J. L. G., and Thomas, H. J. 1987. Influence of the operative conditions on the preparation of pelleted Mo/Al_2O_3 catalysts, *Appl. Catal.* 33: 231–244.

Bolan, N. S., and Barrow, N. J. 1984. Modelling the effect of adsorption of phosphate and other anions on the surface charge of variable charge oxides, *J. Soil Sci.* 35: 273–281.

Bonivardi, A. L., and Baltanas, M. A. 1990. Preparation of Pd/SiO_2 catalysts for methanol synthesis, *J. Catal.* 125: 243–259.

Borchert, A., and Buchholz, K. 1979. Inhomogeneous distribution of fixed enzymes inside carriers, *Biotechnol. Lett.* 1: 15–20.

Borchert, A., and Buchholz, K. 1984. Improved biocatalyst effectiveness by controlled immobilization of enzymes, *Biotechnol. and Bioeng.* 26: 727–736.

Bothe, N. 1982. Struktur, Stabilität und katalytische Aktivität sulfonsaurer Styrol-Divinylbenzol-Harze, PhD Thesis, Technische Universität zu Braunschweig.

Bournonville, J. P., Franck, J. P., and Martino, G. 1983. Influence of the various activation steps on the dispersion and the catalytic properties of platinum supported on chlorinated alumina. In *Preparation of Catalysts III*, eds. Poncelet, G., Grange P., and Jacobs, P. A., pp. 81–90. Amsterdam: Elsevier.

Bowden, J. W., Posner, A. M., and Quirk, J. P. 1977. Ionic adsorption on variable charge mineral surfaces. Theoretical charge development and titration curves, *Aust. J. Soil. Res.* 15: 121–136.

Bowden, J. W., Nagarajah, S., Barrow, N. J., Posner, A. M., and Quirk, J. P. 1980. Describing the adsorption of phosphate, citrate and selenite on a variable-charge mineral surface, *Aust. J. Soil. Res.* 18: 49–60.

Brunelle, J. P. 1978. Preparation of catalysts by adsorption of metal complexes on mineral oxides, *Pure Appl. Chem.* 50: 1211–1229.

Brunelle, J. P., Sugler, A., and Le Page, J.-F. 1976. Active centers of platinum silica catalysts in hydrogenolysis and isomerization of *n*-pentane, *J. Catal.* 43: 273–291.

Brunovska, A. 1987. Dynamic behaviour of a catalyst pellet with nonuniform activity distribution, *Chem. Eng. Sci.* 42: 1969–1976.

Brunovska, A. 1988. Dynamic behaviour of catalyst pellet with step function activity distribution, *Chem. Eng. Sci.* 43: 2546–2548.

Brunovska, A., Morbidelli, M., and Brunovsky, P. 1990. Optimal catalyst pellet activity distributions for deactivating systems, *Chem. Eng. Sci.* 45: 917–925.

Brunovska, A., Remiarova, B., and Pranda, P. 1994. Role of catalyst pellet activity distribution in catalyst poisoning, *Appl. Catal. A* 108: 141–156.

Buchholz, K. 1979. Nonuniform enzyme distribution inside carriers, *Biotechnol. Lett.* 1: 451–456.

Buchholz, K., and Klein, J. 1987. Characterization of immobilized biocatalysts, *Methods Enzymol.* 135: 3–30.

Burch, R., and Hayes, M. J. 1997. The preparation and characterisation of Fe-promoted Al_2O_3-supported Rh catalysts for the selective production of ethanol from syngas, *J. Catal.* 165: 249–261.

Butt, J. B., Downing, D. M., and Lee, J. W. 1977. Inter-intraphase temperature gradients in fresh and deactivated catalyst particles, *Ind. Eng. Chem. Fundam.* 16: 270–272.

Caldwell, A. D., and Calderbank, P. H. 1969. Catalyst dilution – a means of temperature control in packed tubular reactors, *Brit. Chem. Eng.* 14: 1199–1201.

Cannon, K. C., and Hacskaylo, J. J. 1992. Evaluation of palladium impregnation on the performance of a Vycor glass catalytic membrane reactor, *J. Membr. Sci.* 65: 259–268.

Carballo, L., Serrano, C., Wolf, E. E., and Carberry, J. J. 1978. Hydrogen chemisorption studies on supported platinum using the flow technique, *J. Catal.* 52: 507–514.

Carberry, J. J. 1976. *Chemical and Catalytic Reaction Engineering*, New York: McGraw-Hill.

Carleysmith, S. W., Dunnill, P., and Lilly, M. D. 1980a. Kinetic behavior of immobilized penicillin acylase, *Biotechnol. and Bioeng.* 22: 735–756.

Carleysmith, S. W., Eames, M. B. L., and Lilly, M. D. 1980b. Staining method for determination of the penetration of immobilised enzyme into a porous support, *Biotechnol. and Bioeng.* 22: 957–967.

Castro, A. A., Scelza, O. A., Benvenuto, E. R., Baronetti, G. T., De Miguel, S. R., and Parera, J. M. 1983. Competitive adsorption of H_2PtCl_6 and HCl on Al_2O_3 in the preparation of naptha reforming catalysts. In *Preparation of Catalysts III*, eds. Poncelet, G., Grange, P., and Jacobs, pp. 47–56. Amsterdam: Elsevier.

Cervello, J., Hermana, E., Jimenez, J. F., and Melo, F. 1976. Effect of the impregnation conditions on the internal distribution of the active species in catalysts. In *Preparation of Catalysts I*, eds. Delmon, B., Jacobs, P. A., and Poncelet G., pp. 251–263. Amsterdam: Elsevier.

Cervello, J., Melendo, J. F. J., and Hermana, E. 1977. Effect of variable specific rate contant in nonuniform catalysts, *Chem. Eng. Sci.* 32: 155–159.

Charmas, R., Piasecki, W., and Rudzinski, W. 1995. Four layer complexation model for ion adsorption at electrolyte/oxide interface: Theoretical foundations, *Langmuir* 11: 3199–3210.

Chary, K. V. R., Rama Rao, B., and Subrahmanyam, V. S. 1991. Characterization of supported vanadium oxide catalysts by a low-temperature oxygen chemisorption technique III. The V_2O_5/ZrO_2 system, *Appl. Catal.* 74: 1–13.

Che, M., and Bennett, C. O. 1989. The influence of particle size on the catalytic properties of supported metals, *Adv. Catal.* 36: 55–172.

Che, M., and Bonneviot, L. 1988. The change of properties of transition metal ions and the role of support as a function of catalyst preparation. In *Successful Design of Catalysts*, ed. Inui, T., pp. 147–158. Amsterdam: Elsevier.

Chee, Y. C., and Ihm, S. K. 1986. The influence of acid site distribution on the catalytic deactivation of sulfonated poly(styrene–divinylbenzene) membrane catalyst, *J. Catal.* 102: 180–189.

Chemburkar, R. M. 1987. Optimal catalyst activity profiles in pellets. Single pellet theory and experiments, PhD Thesis, University of Notre Dame.

Chemburkar, R. M., Morbidelli, M., and Varma, A. 1987. Optimal catalyst activity profiles in pellets – VII. The case of arbitrary reaction kinetics with finite external heat and mass transport resistances, *Chem. Eng. Sci.* 42: 2621–2632.

Chen, H.-C., and Anderson, R. B. 1973. Study of impregnated chromia on alumina catalysts with an electron probe microanalyzer, *Ind. Eng. Chem. Prod. Res. Dev.* 12: 122–127.

Chen, H.-C., and Anderson, R. B. 1976. Concentration profiles in impregnated chromium and coppper on alumina, *J. Catal.* 43: 200–206.

Chen, H.-C., Gillies, G. C., and Anderson, R. B. 1980. Impregnating chromium and copper in alumina, *J. Catal.* 62: 367–373.

Cheng, W. C., and Pereira, C. J. 1987. Preparation of cobalt–molybdenum hydrotreating catalysts using hydrogen peroxide stabilization, *Appl. Catal.* 33: 331–341.

Chiang, C.-L., and Fang, Z.-R. 1994. Simulation and optimal design for the residual oil hydrodemetallation in a cocurrent moving-bed reactor, *Chem. Eng. Sci.* 49: 1175–1183.

Chiang, C.-L., and Tiou, H.-H. 1992. Optimal design for the residual oil hydrodemetallation in a fixed bed reactor, *Chem. Eng. Commun.* 117: 383–399.

Chibata, I., Tosa, T., and Shibatani, T. 1992. The industrial production of optically active compounds by immobolized biocatalysts. In *Chirality in Industry*, eds. Collins, A. N., Sheldrake, G. N., and Crosby, J., pp. 352–370. Wiley.

Chu, P., Petersen, E. E., and Radke, C. J. 1989. Modeling wet impregnation of nickel on γ-alumina, *J. Catal.* 117: 52–70.

Chung, B. H., and Chang, H. N. 1986. Effect of internal diffusion on the apparent stability of nonuniformly distributed biocatalysts, *Korean J. Chem. Eng.* 3: 39–43.

Clause, O., Kermarec, M., Bonneviot, L., Villain, F., and Che, M. 1992. Nickel(II) ion-support interactions as a function of preparation method of silica-supported nickel materials, *J. Am. Chem. Soc.* 114: 4709–4717.

Contescu, C., and Vass, M. I. 1987. The effect of pH on the adsorption of palladium(II) complexes on alumina, *Appl. Catal.* 33: 259–271.

Contescu, C., Hu, J., and Schwarz, J. A. 1993a. 1-pK multisites description of charge development at the aqueous alumina interface, *J. Chem. Soc. Faraday Trans.* 89: 4091–4099.

Contescu, C., Jagiello, J., and Schwarz, J. A. 1993b. Heterogeneity of proton binding sites at the oxide/solution interface, *Langmuir* 9: 1754–1765.

Contescu, C., Macovei, D., Craiu, C., Teodorescu, C., and Schwarz, J. A. 1995a. Thermal induced evolution of chlorine-containing precursors in impregnated Pd/Al_2O_3 catalysts, *Langmuir* 11: 2031–2040.

Contescu, C., Jagiello, J., and Schwarz, J. A. 1995b. Proton affinity distributions: A scientific basis for the design and construction of supported metal catalysts. In *Preparation of Catalysts VI*, eds. Poncelet, G., Martens, J., Delmon, B., Jacobs, P. A., and Grange, P., pp. 237–252. Amsterdam: Elsevier.

Corbett, W. E., and Luss, D. 1974. The influence of non-uniform catalytic activity on the performance of a single spherical pellet, *Chem. Eng. Sci.* 29: 1473–1483.

Cot, L., Guizard, C., and Larbot, A. 1988. Novel ceramic material for liquid separation process: Present and prospective applications in microfiltration and ultrafiltration, *Ind. Ceram.* 8: 143–148.

Coulson, J. M., and Richardson, J. F. 1991. Drying. In *Chemical Engineering*, 4th ed., vol. 2, Chapter 16. Oxford: Pergamon.

Cukierman, A. L., Laborde, M. A., and Lemcoff, N. O. 1983. Optimum activity distribution in a catalyst pellet for a complex reaction, *Chem. Eng. Sci.* 38: 1977–1982.

Cusumano, J. A. 1991. Creating the future of the chemical industry – catalysts by molecular design. In *Perspectives in Catalysis*, eds. Thomas, J. M., and Zamaraev, K. I., pp. 1–33. Oxford: Blackwell Scientific.

Dadyburjor, D. B. 1982. Distribution for maximum activity of a composite catalyst, *A.I.Ch.E.J.* 28: 720–728.

Dadyburjor, D. B. 1985. Selectivity over unifunctional multicomponent catalysts with nonuniform distribution of components, *Ind. Eng. Chem. Fundam.* 24: 16–27.

Dadyburjor, D. B., and White, C. W., III 1990. Effect of position-dependent deactivation on the design of a composite cracking catalyst, *Chem. Eng. Sci.* 45: 2619–2624.

D'Aniello, M. J. 1981. Anion adsorption on alumina, *J. Catal.* 69: 9–17.

Davis, J. A., and Leckie, J. O. 1978. Surface ionization and complexation at the oxide/water interface 2. Surface properties of amorphous iron oxyhydroxide and adsorption of metal ions, *J. Coll. Interf. Sci.* 67: 90–107.

Davis, J. A., and Leckie, J. O. 1980. Surface ionization and complexation at the oxide/water interface 3. Adsorption of anions, *J. Coll. Interf. Sci.* 74: 32–43.

Davis, J. A., James, R. O., and Leckie J. O. 1978. Surface ionization and complexation at the oxide/water interface 1. Computation of electrical double layer properties in simple electrolytes, *J. Coll. Interf. Sci.* 63: 480–499.

De Jong, K. P. 1991. Deposition precipitation onto pre-shaped carrier bodies. Possibilities and limitations. In *Preparation of Catalysts V*, eds. Poncelet, G., Jacobs, P. A., Grange, P., and Delmon, B., pp. 19–36. Amsterdam: Elsevier.

DeLancey, G. B. 1973. An optimal catalyst activation policy for poisoning problems, *Chem. Eng. Sci.* 28: 105–118.

Delannay, F. (ed.) 1984. *Characterization of Heterogeneous Catalysts*. New York: Marcel Dekker.

Delmon, B., and Houalla, M. 1979. Tentative classification of the factors influencing the reduction step in the activation of supported catalysts. In *Preparation of Catalysts II*, eds. Delmon, B., Grange, P., Jacobs, P., and Poncelet, G., pp. 439–468. Amsterdam: Elsevier.

De Miguel, S. R., Scelza, O. A., Castro, A. A., Baronetti, G. T., Ardiles, D. R., and Parera, J. M. 1984. Radial profiles in Pt/Al_2O_3, Re/Al_2O_3, and Pt–Re/Al_2O_3, *Appl. Catal.* 9: 309–315.

D'Espinose de la Caillerie, J.-B., Bobin, C., Rebours, B., and Clause, O. 1995. Alumina/water interfacial phenomena during impregnation. In *Preparation of Catalysts VI*, eds. Poncelet, G., Martens, J., Delmon, B., Jacobs, P., and Grange, P., pp. 169–184. Amsterdam: Elsevier.

Do, D. D. 1984. A method for solving diffusion and reaction problems with nonuniform activity catalysts, *Chem. Eng. Sci.* 39: 1519–1522.

Do, D. D. 1985. Modeling of impregnation kinetics and internal activity profiles: Adsorption of HCl, $HReO_4$ and H_2PtCl_6 onto γ-Al_2O_3. *Chem. Eng. Sci.* 40: 1871–1880.

Do, D. D., and Bailey, J. E. 1982. Approximate analytical solutions for porous catalysts with nonuniform activity, *Chem. Eng. Sci.* 37: 545–551.

Dorling, T. A., Lynch, B. W. J., and Moss, R. L. 1971. The structure and activity of supported metal catalysts. V. Variables in the preparation of platinum/silica catalysts, *J. Catal.* 20: 190–201.

Dorling, T. A., and Moss, R. L. 1967. The structure and activity of supported metal catalysts. II. Crystallite size and CO chemisorption in platinum/silica catalysts, *J. Catal.* 7: 378–385.

Dougherty, R. C., and Verykios, X. E. 1987. Nonuniformly activated catalysts, *Catal. Rev. – Sci. Eng.* 29: 101–150.

Downing, D. M., Lee, J. W., and Butt, J. B. 1979. Simulation models for intraparticle deactivation: Scope and reliability, *A.I.Ch.E.J.* 25: 461–469.

Duncombe, P. R., and Weller, S. W. 1985. Thermal behavior of shell molybdena–alumina catalysts, *A.I.Ch.E.J.* 31: 410–414.

Engels, S., Lausch, H., and Schwokowski, R. 1987. Untersuchungen zur Adsorption bi- und polyfunktioneller organischer Saeuren an Aluminiumoxid, *Chem. Techn. (Leipzig)* 39: 387–391.

Ernst, W. R., and Daugherty, D. J. 1978. Method for the study of performance of a single spherical particle with nonuniform catalytic activity, *A.I.Ch.E.J.* 24: 935–937.

Fenelonov, V. B. 1975. Analysis of the drying stages in the technology of deposited catalysts I. Role of the pore structure and the drying conditions, *Kinet. and Catal.* 16: 628–635.

Fenelonov, V. B., Neimark, A. V., and Kheifets, L. I. 1979. Analysis of steps of impregnation and drying in preparation of supported catalysts. In *Preparation of Catalysts II*, eds. Delmon, B., Grange, P., Jacobs, P., and Poncelet, G., pp. 233–244. Amsterdam: Elsevier.

Finlayson, B. A. 1980. *Nonlinear Analysis in Chemical Engineering*. New York: McGraw-Hill.

Foger, K. 1984. Dispersed metal catalysts. In *Catalysis Science and Technology*, vol. 6, eds. Anderson, J. R., and Boudart, M., pp. 227–305. Berlin: Springer-Verlag.

Frost, A. C., Sawyer, J. E., Summers, J. C., Shah, Y. T., and Dassori, C. G. 1990. Reactor model for the fixed bed catalytic incineration of VOC's, presented at the 83rd Annual Meeting and Exhibition, Air and Waste Management Association, Pittsburgh, PA, June 24–29.

Fuentes, S., and Figueras, F. 1980. The influence of particle size on the catalytic properties of alumina-supported rhodium catalysts, *J. Catal.* 61: 443–453.

Fujiyama, T., Otsuka, M., Tsuki, H., and Ueno, A. 1987. Control of the impregnation profile of Ni in an Al_2O_3 sphere, *J. Catal.* 104: 323–330.

Furusaki, S. 1988. Engineering aspects of immobilised biocatalysts, *J. Chem. Eng. Japan* 21: 219–230.

Galiasso, R., de Ochoa, O. L., and Andreu, P. 1983. Influence of the drying rate on the distribution of active metals and on the activity of heavy crude hydrotreating catalyst, *Appl. Catal.* 5: 309–322.

Gates, G. C., Winouskas, J. S., and Heath, H. W. 1972. The dehydration of t-butyl alcohol catalyzed by sulfonic acid resin, *J. Catal.* 24: 320–327.

Gavriilidis, A., and Varma, A. 1992. Optimal catalyst activity profiles in pellets: 9. Study of ethylene epoxidation, *A.I.Ch.E.J.* 38: 291–296.

Gavriilidis, A., Varma, A., and Morbidelli, M. 1993a. Optimal distribution of catalyst in pellets, *Catal. Rev. – Sci. Eng.* 35: 399–456.

Gavriilidis, A., Sinno, B., and Varma, A. 1993b. Influence of loading on metal surface area for Ag/α-Al_2O_3 catalysts, *J. Catal.* 139: 41–47.

Gazimzyanov, N. R., Mikhailov, V. I., and Zadko, I. I. 1992. Study of distribution of active components in alumino–nickel–molybdenum catalysts obtained by a single impregnation, *Kinet. and Catal.* 33: 737–743.

Gelosa, D., Pedeferri, M. P., and Morbidelli, M. 1998. Esterification reactions with nonuniformly sulfonated resins, First Annual Report INTREASEP Project, Joule Contract JOE3-CT97-0082.

Gladden, L. F. 1995. Applications of nuclear magnetic resonance imaging in particle technology, *Part. Part. Syst. Charact.* 12: 59–67.

Gottifredi, J. C., Gonzo, E. E., and Quiroga, O. D. 1981. Isothermal effectiveness factor II. Analytical expression for single reaction with arbitrary kinetics, geometry and activity distribution, *Chem. Eng. Sci.* 36: 713–719.

Goula, M. A., Kordulis, C., Lycourghiotis, A., and Fierro, J. L. G. 1992a. Development of molybdena catalysts supported on γ-alumina extrudates with four different Mo profiles: Preparation, characterisation and catalytic properties, *J. Catal.* 137: 285–305.

Goula, M. A., Kordulis, C., and Lycourghiotis, A. 1992b. Influence of impregnation parameters on the axial Mo/γ-alumina profiles studied using a novel simple technique, *J. Catal.* 133: 486–497.

Guisan, J. M., Serrano, J., Melo, F. V., and Ballesteros, A. 1987. Mixed enzymic reaction–internal diffusion kinetics of nonuniformly distributed immobilised enzymes. The system agarose–micrococcal endonuclease, *Appl. Biochem. Biotech.* 14: 49–72.

Hachiya, K., Sasaki, M., Saruta, Y., Mikami, N., and Yasunaga, T. 1984. Static and kinetic studies of adsorption–desorption of metal ions on a γ-Al_2O_3 surface. 1. Static study of adsorption–desorption, *J. Phys. Chem.* 88: 23–27.

Hamada, H. Funaki, R., Kuwahara, Y., Kintaichi, Y., Wakabayashi, K., and Ito, T. 1987. Systematic preparation of supported Rh catalysts having desired metal particle size by using silica supports with controlled pore structure, *Appl. Catal.* 30: 177–180.

Hanika, J., and Ehlova, V. 1989. Modelling of internal diffusion inside a catalyst particle with non-uniform radial activity profile for parallel reactions, *Collect. Czech. Chem. Commun.* 54: 81–90.

Hanika, J., Machek, V., Nemec, V., Ruzicka, V., and Kunz, J. 1982. Simultaneous diffusion and adsorption of chloroplatinic acid in charcoal pellet during preparation process of supported platinum catalyst, *J. Catal.* 77: 248–256.

Hanika, J., Hao, L. H., and Ruzicka, V. 1983. Relation between the conditions of preparation and the activity of supported platinum catalysts, *Collect. Czech. Chem. Commun.* 48: 3079–3085.

Hanika, J., Janousek, V., and Sporka, K. 1987a. Modelling of impregnation of γ-alumina with cobalt and molybdenum salts. $CoCl_2-(NH_4)_2MoO_4-\gamma-Al_2O_3$ aluminate type system, *Collect. Czech. Chem. Commun.* 52: 663–671.

Hanika, J., Janousek, V., and Sporka, K. 1987b. Modelling of impregnation of γ-alumina with cobalt and molybdenum salts. $Co(NO_3)_2-(NH_4)_2MoO_4-\gamma-Al_2O_3$ chloride type system, *Collect. Czech. Chem. Commun.* 52: 672–677.

Harold, M. P., and Luss, D. 1987. Impact of the catalytic activity profile on observed multiplicity features: CO oxidation on Pt/Al_2O_3, *Ind. Eng. Chem. Res.* 26: 1616–1621.

Harriott, P. 1969. Diffusion effects in the preparation of impregnated catalysts, *J. Catal.* 14: 43–48.

Haworth, A. 1990. A review of the modelling of sorption from aqueous solution, *Adv. Coll. Interf. Sci.* 32: 43–78.

Hayes, K. F., and Leckie, J. O. 1986. Mechanism of lead ion adsorption at the goethite–water interface, *ACS Symp. Ser.* 323: 114–141.

Heck, R. M., and Farrauto, R. J. 1995. *Catalytic Air Pollution Control. Commercial Technology*, New York: Van Nostrand Reinhold.

Hegedus, L. L. 1980. Catalyst pore structures by constrained nonlinear optimisation, *Ind. Eng. Chem. Prod. Res. Dev.* 19: 533–537.

Hegedus, L. L., and Gumbleton, J. J. 1980. Catalysts, computers and cars: A growing symbiosis, *Chemtech*, October, 630–642.

Hegedus, L. L., and McCabe, R. 1984. *Catalyst Poisoning*. New York: Marcel Dekker.

Hegedus, L. L., and Pereira, C. J. 1990. Reaction engineering for catalyst design, *Chem. Eng. Sci.* 45: 2027–2044.

Hegedus, L. L., and Summers, J. C. 1977. Improving the poison resistance of supported catalysts, *J. Catal.* 48: 345–353.

Hegedus, L. L., Summers, J. C., Schlatter, J. C., and Baron, K. 1979a. Poison resistant catalysts for the simultaneous control of hydrocarbon, carbon monoxide and nitrogen oxide emissions, *J. Catal.* 56: 321–335.

Hegedus, L. L., Chou, T. S., Summers, J. C., and Potter, N. M. 1979b. Multicomponent chromatographic processes during the impregnation of alumina pellets with noble metals. In *Preparation of Catalysts II*, eds. Delmon, B., Grange, P., Jacobs, P., and Poncelet, G., pp. 171–184. Amsterdam: Elsevier.

Heise, M. S., and Schwarz, J. A. 1985. Preparation of metal distributions within catalyst supports I. Effect of pH on catalytic metal profiles, *J. Coll. Interf. Sci.* 107: 237–243.

Heise, M. S., and Schwarz, J. A. 1986. Preparation of metal distributions within catalyst supports II. Effect of ionic strength on catalytic metal profiles, *J. Coll. Interf. Sci.* 113: 55–61.

Heise, M. S., and Schwarz, J. A. 1987. Preparation of metal distributions within catalyst supports. In *Preparation of Catalysts IV*, eds. Delmon, B., Grange, P., Jacobs, P. A., and Poncelet, G., pp. 1–13. Amsterdam: Elsevier.

Heise, M. S., and Schwarz, J. A. 1988. Preparation of metal distributions within catalyst supports III. Single component modeling of pH, ionic strength and concentration effects, *J. Coll. Interf. Sci.* 123: 51–58.

Hepburn, J. S., Stenger, H. G., and Lyman, C. E. 1989a. Distributions of HF co-impregnated rhodium, platinum and palladium in alumina honeycomb supports, *Appl. Catal.* 55: 271–285.

Hepburn, J. S., Stenger, H. G., and Lyman, C. E. 1989b. Effects of drying on the preparation of HF co-impregnated rhodium/Al_2O_3 catalysts, *Appl. Catal.* 55: 287–299.

Hepburn, J. S., Stenger, H. G., and Lyman, C. E. 1989c. Co-impregnation of rhodium into alumina honeycombs with acids and salts, *Appl. Catal.* 56: 107–118.

Hepburn, J. S., Stenger, H. G., and Lyman, C. E. 1991a. Co-impregnated Rh/Al₂O₃. I. Preparation, *J. Catal.* 128: 34–47.

Hepburn, J. S., Stenger, H. G., and Lyman, C. E. 1991b. Co-impregnated Rh/Al₂O₃. II. Nitric oxide reduction and SO₂ poisoning, *J. Catal.* 128: 48–62.

Hepburn, J. S., Stenger, H. G., and Lyman, C. E. 1991c. Co-impregnation of rhodium chloride with hydrofluoric acid into dry and pre-wet alumina, *Appl. Catal.* 71: 205–218.

Hiemstra, T., van Riemsdijk, W. H., and Bolt, G. H. 1989a. Multisite proton adsorption modeling at the solid/solution interface of hydroxides: A new approach. I. Model description and evaluation of intrinsic reaction constants, *J. Coll. Interf. Sci.* 133: 91–104.

Hiemstra, T., de Wit, J. C. M., and van Riemsdijk, W. H. 1989b. Multisite proton adsorption modeling at the solid/solution interface of hydroxides: A new approach. II. Application to various important hydroxides, *J. Coll. Interf. Sci.* 133: 105–117.

Hoekstra, J. 1968. Catalyst for treatment of combustible waste products, U.S. Patent 3,388,077.

Hohl, H., and Stumm, W. 1976. Interaction of Pb^{2+} with hydrous γ-Al₂O₃, *J. Coll. Interf. Sci.* 55: 281–288.

Hollewand, M. P., and Gladden, L. F. 1992. Visualization of phases in catalyst pellets and pellet mass transfer processes using magnetic resonance imaging, *Trans. IChemE* 70A: 183–185.

Hollewand, M. P., and Gladden, L. F. 1994. Probing the structure of porous pellets: An NMR study of drying, *Magn. Reson. Imaging*, 12, 291–294.

Horvath, C., and Engasser, J.-M. 1973. Pellicular heterogeneous catalysts. A theoretical study of the advantages of shell structured immobilized enzyme particles, *Ind. Eng. Chem. Fundam.* 12: 229–235.

Hossain, M. M., and Do, D. D. 1987. Effects of nonuniform immobilised enzyme distribution in porous solid supports on the performance of a continuous reactor, *Chem. Eng. J.* 34: B35–B47.

Hsieh, H. P. 1991. Inorganic membrane reactors, *Catal. Rev. Sci. Eng.* 33: 1–70.

Hu, Z., Wan, C. Z., Lui, Y. K., Dettling, J., and Steger, J. J. 1996. Design of a novel Pd three-way catalyst: Integration of catalytic functions in three dimensions, *Catal. Today* 30: 83–89.

Huang, C. P. 1975. Adsorption of phosphate at the hydrous γ-Al₂O₃–electrolyte interface, *J. Coll. Interf. Sci.* 53: 178–186.

Huang, Y.-J. R., Barrett, B. T., and Schwarz, J. A. 1986. The effect of solution variables on metal weight loading during catalyst preparation, *Appl. Catal.* 24: 241–248.

Huang, Y.-J., and Schwarz, J. A. 1987. The effect of catalyst preparation on catalytic activity: III. The catalytic activity of Ni/Al₂O₃ catalysts prepared by incipient wetness, *Appl. Catal.* 32: 45–57.

Iannibello, A., and Mitchell, P. C. H. 1979. Preparative chemistry of cobalt–molybnenum/ alumina catalysts. In *Preparation of Catalysts II*, eds. Delmon, B., Grange, P., Jacobs, P., and Poncelet, G., pp. 469–478. Amsterdam: Elsevier.

Iannibello, A., Marengo, S., Trifiro, F., and Villa, P. L. 1979. A study of the chemisorption of chromium(VI), molybdenum(VI) and tungsten(VI) onto γ-alumina. In *Preparation of Catalysts II*, eds. Delmon, B., Grange, P., Jacobs, P., and Poncelet, G., pp. 65–76. Amsterdam: Elsevier.

Iglesia, E. 1997. Design, synthesis, and use of cobalt-based Fischer-Tropsch synthesis catalysts, *Appl. Catal. A.* 161: 59–78.

Iglesia, E., Reyes, S. C., Madon, R. J., and Soled, S. L. 1993. Selectivity control and catalyst design in the Fischer–Tropsch synthesis: Sites, pellets and reactors, *Adv. Catal.* 39: 221–302.

Iglesia, E., Soled, S. L., Baumgartner, J. E., and Reyes, S. C. 1995. Synthesis and catalytic properties of eggshell cobalt catalysts for the Fischer–Tropsch synthesis, *J. Catal.* 153: 108–122.

Jianguo, W., Jiayu, Z., and Li, P. 1983. The role of competitive adsorbate in the impregnation of platinum in pelleted alumina support. In *Preparation of Catalysts III*, eds. Poncelet, G., Grange P., and Jacobs, P. A., pp. 57–67. Amsterdam: Elsevier.

Jiratova, K. 1981. Isoelectric point of modified alumina, *Appl. Catal.* 1: 165–167.

Johnson, D. L., and Verykios, X. E. 1983. Selectivity enhancement in ethylene oxidation employing partially impregnated catalysts, *J. Catal.* 79: 156–163.

Johnson, D. L., and Verykios, X. E. 1984. Effects of radially nonuniform distributions of catalytic activity on performance of spherical catalyst pellets, *A.I.Ch.E.J.* 30: 44–50.

Juang, H.-D., and Weng, H.-S. 1983. Performance of catalysts with nonuniform activity profiles 2. Theoretical analysis for nonisothermal reactions. *Ind. Eng. Chem. Fundam.* 22: 224–230.

Juang, H.-D., and Weng, H.-S. 1984. Performance of biocatalysts with nonuniformly distributed immobilized enzyme, *Biotechnol. and Bioeng.* 26, 623–626.

Juang, H.-D., Weng, H.-S. and Wang, C.-C. 1981. The performance of catalysts with nonuniform activity profile I. Theoretical analysis for isothermal reactions, *Stud. Surf. Sci. Catal.* 7: 866–876.

Karakonstantis, L., Kordulis, Ch., and Lycourghiotis, A. 1992. Mechanism of adsorption of tungstates on the interface of γ-alumina/electrolyte solutions, *Langmuir* 8: 1318–1324.

Karpe, P., and Ruckenstein, E. 1989. Role of electrokinetic phenomena in the preparation of supported metal catalysts, *Coll. and Polym. Sci.* 267: 145–150.

Kasaoka, S., and Sakata, Y. 1968. Effectiveness factors for non-uniform catalyst pellets, *J. Chem. Eng. Japan* 1: 138–142.

Kehoe, P. 1974. Letter to the Editors, *Chem. Eng. Sci.* 29: 315.

Kehoe, J. P. G., and Butt, J. B. 1972. Interactions of inter- and intraphase gradients in a diffusion limited catalytic reaction, *A.I.Ch.E.J.* 18: 347–355.

Keil, F. J., and Rieckmann, C. 1994. Optimisation of three-dimensional catalyst pore structures, *Chem. Eng. Sci.* 49: 4811–4822.

Keller, J. B., Falkovitz, M. S., and Frisch, H. 1984. Optimal catalyst distribution in a membrane, *Chem. Eng. Sci.* 39: 601–604.

Kempling, J. C., and Anderson, R. B. 1970. Hydrogenolysis of *n*-butane on supported ruthenium, *Ind. Eng. Chem. Proc. Des. Dev.* 9: 116–120.

Kheifets, L. I., Neimark, A. V., and Fenelonov, V. B. 1979. Analysis of the influence of permeation and drying stages in supported catalyst preparation on the distribution of components I. Precipitation of one component from solution, *Kinet. and Catal.* 20: 626–632.

Kim, D. S., Kurusu, Y., Wachs, I. E., Hardcastle, F. D., and Segawa, K. 1989. Physicochemical properties of MoO_3–TiO_2 prepared by an equilibrium adsorption method, *J. Catal.* 120: 325–336.

Kimura, H. 1993. Selective oxidation of glycerol on a platinum–bismuth catalyst by using a fixed bed reactor, *Appl. Catal. A* 105: 147–158.

King, D. L. 1978. A Fischer–Tropsch study of supported ruthenium catalysts, *J. Catal.* 51: 386–397.

Klein, J., Widdecke, H., and Bothe, N. (1984). Influence of functional group distribution on the thermal stability and catalytic activity of sulfonated styrene-DVB-copolymers, *Macromol. Chem. and Phys., Suppl.* 6: 211–226.

Klugherz, P. D., and Harriott, P. 1971. Kinetics of ethylene oxidation on a supported silver catalyst, *A.I.Ch.E.J.* 17: 856–866.

Ko, S. H., and Chou, T. C. 1994. Hydrogenation of (–)-α-pinene over nickel–phosphorus/aluminum oxide catalysts prepared by electroless deposition, *Can. J. Chem. Eng.* 72: 862–873.

Komiyama, M. 1985. Design and preparation of impregnated catalysts, *Catal. Rev. – Sci. Eng.* 27: 341–372.

Komiyama, M., and Muraki, H. 1990. Design of intrapellet activity profiles in supported catalysts. In *Handbook of Heat and Mass Transfer*, vol. 4, ed. Cheremisinoff, N. P., pp. 447–500. Houston: Gulf Publishing Company.

Komiyama, M., Ohashi, K., Morioka, Y., and Kobayashi, J. 1997. Effects of the intrapellet activity profiles on the selectivities in consecutive reactions: 1,3-butadiene hydrogenation over Pt/Al_2O_3, *Bull. Chem. Soc. Japan* 70: 1009–1013.

Komiyama M Merrill, R, P., and Harnsberger, H. F. 1980. Concentration profiles in impregnation of porous catalysts: Nickel on alumina, *J. Catal.* 63: 35–52.

Kotter, M., and Riekert, L. 1979. The influence of impregnation, drying and activation on the activity and distribution of CuO on α-alumina. In *Preparation of Catalysts II*, eds. Delmon, B., Grange, P., Jacobs, P., and Poncelet, G., pp. 51–63. Amsterdam: Elsevier.

Kotter, M., and Riekert, L. 1983a. Impregnation-type catalysts with nonuniform distribution of the active component. Part I: Influence of the accessibility of the active component on activity and selectivity, *Chem. Eng. Fundam.* 2: 22–30.

Kotter, M., and Riekert, L. 1983b. Impregnation-type catalysts with nonuniform distribution of the active component. Part II: Preparation and properties of catalysts with different distribution of the active component on inert carriers, *Chem. Eng. Fundam.* 2: 31–38.

Kresge, C. T., Chester, A. W., and Oleck, S. M. 1992. Control of metal radial profiles in alumina supports by carbon dioxide, *Appl. Catal. A* 81: 215–226.

Kulkarni, S. S., Mauze. G. R., and Schwarz, J. A. 1981. Concentration profiles and the design of metal-supported catalysts, *J. Catal.* 69: 445–453.

Kummer, J. T. 1980. Catalysts for automobile emission control, *Prog. Energy Combust. Sci.* 6: 177–199.

Kummert, R., and Stumm, W. 1980. The surface complexation of organic acids on hydrous γ-Al_2O_3, *J. Colloid Interface Sci.* 75: 373–385.

Kunimori, K., Nakajima, I., and Uchijima, T. 1982. Catalytic performance of egg white type Pt/Al_2O_3 catalyst in the oxidation of C_3H_8 and CO, *Chem. Lett.* 1165–1168.

Kunimori, K., Kawasaki, E., Nakajima, I., and Uchijima, T. 1986. Catalytic performance of egg-white type Pt/Al_2O_3 catalyst: Multiple steady states in the oxidation of CO, *Appl. Catal.* 22: 115–122.

Lee, C. K., and Varma, A. 1987. Steady state multiplicity behavior of an adiabatic plug-flow reactor with nonuniformly active catalyst, *Chem. Eng. Commun.* 58: 287–309.

Lee, C. K., and Varma, A. 1988. An isothermal fixed-bed reactor with nonuniformly active catalysts: Experiments and theory, *Chem. Eng. Sci.* 43: 1995–2000.

Lee, C. K., Morbidelli, M., and Varma, A. 1987a. Optimal catalyst activity profiles in pellets 6. Optimization of the isothermal fixed-bed reactor with multiple zones, *Ind. Eng. Chem. Res.* 26: 167–170.

Lee, C. K., Morbidelli, M., and Varma, A. 1987b. Steady state multiplicity behavior of an isothermal axial dispersion fixed-bed reactor with nonuniformly active catalyst, *Chem. Eng. Sci.* 42: 1595–1608.

Lee, H. H. 1981. Generalized effectiveness factor for pellets with nonuniform activity distribution: Asymptotic region of strong diffusion effect, *Chem. Eng. Sci.* 36: 1921–1925.

Lee, H. H. 1984. Catalyst preparation by impregnation and activity distribution, *Chem. Eng. Sci.* 39: 859–864.

Lee, H. H. 1985. Catalyst preparation by impregnation and activity distributions – II. Coimpregnation for nonuniform distribution based on competitive adsorption, *Chem. Eng. Sci.* 40: 1295–1300.

Lee, H. H., and Ruckenstein, E. 1983. Catalyst sintering and reactor design, *Cat. Rev. – Sci. Eng.* 25: 475–550.

Lee, J. W., Butt, J. B., and Downing, D. M. 1978. Kinetic, transport and deactivation rate interactions on steady state and transient responses in heterogeneous catalysis, *A.I.Ch.E.J.* 24: 212–222.

Lee, S. H., and Ruckenstein, E. 1986. Optimum design of zeolite/silica–alumina catalysts, *Chem. Eng. Commun.* 46: 43–64.

Lee, S.-Y., and Aris, R. 1985. The distribution of active ingredients in supported catalysts prepared by impregnation, *Catal. Rev. – Sci. Eng.* 27: 207–340.

Le Page, J.-F. 1987. *Applied Heterogeneous Catalysis: Design, Manufacture, Use of Solid Catalysts*, Paris: Editions Technip.

Letkova, Z., Brunovska, A., and Markos, J. 1994. Study of activity distribution for consecutive reactions, *Collect. Czech. Chem. Commun.* 59: 1788–1799.

Li, B., Liang, J., Han, B., and Liu, Y. 1985. The distribution control of active components on support, *J. Catal.* (Chinese) 6: 95–99.

Li, W. D., Li, Y. W., Qin, Z. F., and Chen, S. Y. 1994. Theoretical prediction and experimental validation of the egg-shell distribution of Ni for supported Ni/Al$_2$O$_3$ catalysts, *Chem. Eng. Sci.* 49: 4889–4895.

Lilly, M. D. 1996. Biotransformations using immobilised biocatalysts – past, present and future. In *Advances in Molecular and Cell Biology*, vol. 15A, ed. Danielsson, B., pp. 139–147. London: JAI Press.

Limbach, K. W., and Wei, J. 1988. Effect of nonuniform activity on hydrodemetallation catalyst, *A.I.Ch.E.J.* 34: 305–313.

Lin, T.-B., and Chou, T.-C. 1994. Selective hydrogenation of isoprene on eggshell and uniform palladium profile catalysts, *Appl. Catal. A* 108: 7–19.

Lin, T.-B., and Chou, T.-C. 1995. Pd migration. 1. A possible reason for the deactivation of pyrolysis gasoline partial hydrogenation catalysts, *Ind. Eng. Chem. Res.* 34: 128–134.

Lundquist, E. G. 1995. Catalysed esterification process, U.S. Patent 5,426,199.

Lycourghiotis, A. 1995. Preparation of supported catalysts by equilibrium deposition–filtration. In *Preparation of Catalysts VI*, eds. Poncelet, G., Martens, J., Delmon, B., Jacobs, P. A., and Grange, P., pp. 95–129. Amsterdam: Elsevier.

Lyman, C. E., Hepburn, J. S., and Stenger, H. G. 1990. Quantitative Pt and Rh distributions in pollution–control catalysts, *Ultramicroscopy.* 34: 73–80.

Maatman, R. W. 1959. How to make a more effective platinum–alumina catalyst, *Ind. Eng. Chem.* 51: 913–914.

Maatman, R. W., and Prater, C. D. 1957. Adsorption and exclusion in impregnation of porous catalytic supports, *Ind. Eng. Chem.* 49: 253–257.

Machek, V., Hanika, J., Sporka, K., Ruzicka, V., Kunz, J., and Janacek, L. 1983a. The influence of solvent nature of chloroplatinic acid used for support impregnation on the distribution, dispersity and activity of platinum hydrogenation catalysts. In *Preparation of Catalysts III*, eds. Poncelet, G., Grange, P., and Jacobs, P. A., pp. 69–80. Amsterdam: Elsevier.

Machek, V., Ruzicka, V., Sourkova, M., Kunz, J., and Janacek, L. 1983b. Preparation of Pt/activated carbon and Pt/alumina catalysts by impregnation with platinum complexes, *Collect. Czech. Chem. Commun.* 48: 517–526.

Maitra, A. M., Cant, N. W., and Trimm, D. L. 1986. The preparation of tungsten based catalysts by impregnation of alumina pellets, *Appl. Catal.* 27: 9–19.

Mang, Th., Breitscheidel, B., Polanek, P., and Knoezinger, H. 1993. Adsorption of platinum complexes on silica and alumina: Preparation of non-uniform metal distributions within support pellets, *Appl. Catal. A* 106: 239–258.

Markos, J., and Brunovska, A. 1989. Optimal catalyst activity distribution in fixed-bed reactor with catalyst deactivation, *Collect. Czech. Chem. Commun.* 54: 375–387.

Markos, J., Brunovska, A., and Letkova, Z. 1990. Optimal catalyst pellet activity distributions. Fixed-bed reactor with catalyst deactivation, *Comput. Chem. Eng.* 14: 1317–1322.

Mars, P., and Gorgels, M. J. 1964. Hydrogenation of acetylene – a theory of selectivity. In *Chemical Reaction Engineering: Proceedings of the Third European Symposium*, Supplement to *Chem. Eng. Sci.*, pp. 55–65. Oxford: Pergamon Press.

Martin, G. R., White, C. W., III, and Dadyburjor, D. B. 1987. Design of zeolite/silica–alumina catalysts for triangular cracking reactions, *J. Catal.* 106: 116–124.

Masi, M., Sangalli, M., Carra, S., Cao, G., and Morbidelli, M. 1988. Kinetics of ethylene hydrogenation on supported platinum. Analysis of multiplicity and nonuniformly active catalyst particle behavior, *Chem. Eng. Sci.* 43: 1849–1854.

Mathur, I., Bakhshi, N. N., and Mathews, J. F. 1972. Effect of ultrasonic waves on the activity of chromia–alumina catalyst *Can. J. Chem. Eng.* 50: 344–348.

McCoy, M. 1999. Catalyst makers look for growth, *Chemical and Engineering News*, September 1999, pp. 17–25.

Melis, S., Varma, A., and Pereira, C. J. 1997. Optimal distribution of catalyst for a case involving heterogeneous and homogeneous reactions, *Chem. Eng. Sci.* 52: 165–169.

Melo, F., Cervello, J., and Hermana, E. 1980a. Impregnation of porous supports – I. Theoretical study of the impregnation of one or two active species, *Chem. Eng. Sci.* 35: 2165–2174.

Melo, F., Cervello, J., and Hermana, E. 1980b. Impregnation of porous supports – II. System Ni/Ba on alumina, *Chem. Eng. Sci.* 35: 2175–2184.

Michalko, E. 1966a. Method for oxidizing gaseous combustible waste products, U.S. Patent 3,259,454.

Michalko, E. 1966b. Preparation of catalyst for the treatment of combustible waste products, U.S. Patent 3,259,589.

Mieth, J. A., Schwarz, J. A., Huang, Y.-J., and Fung, S. C. 1990. The effect of chloride on the point of zero charge of γ-Al_2O_3, *J. Catal.* 122: 202–205.

Minhas, S., and Carberry, J. J. 1969. On the merits of partially impregnated catalysts, *J. Catal.* 14: 270–272.

Morbidelli, M., and Varma, A. 1982. Optimal catalyst activity profiles in pellets 2. The influence of external mass transfer resistance, *Ind. Eng. Chem. Fundam.* 21: 284–289.

Morbidelli, M., and Varma, A. 1983. On shape normalization for non-uniformly active catalyst pellets II, *Chem. Eng. Sci.* 38: 297–305.

Morbidelli, M., Servida, A., and Varma, A. 1982. Optimal catalyst activity profiles in pellets 1. The case of negligible external mass transfer resistance, *Ind. Eng. Chem. Fundam.* 21: 278–284.

Morbidelli, M., Servida, A., Paludetto, R., and Carra, S. 1984a. Optimal catalyst design for ethylene oxide synthesis, *J. Catal.* 87: 116–125.

Morbidelli, M., Servida, A., and Varma, A. 1984b. Optimal distribution of immobilized enzyme in a pellet for a substrate-inhibited reaction, *Biotechnol. and Bioeng.* 26: 1508–1510.

Morbidelli, M., Servida, A., Carra, S., and Varma, A., 1985. Optimal catalyst activity profiles in pellets 3. The nonisothermal case with negligible external transport limitations, *Ind. Eng. Chem. Fundam.* 24: 116–119.

Morbidelli, M., Servida, A., and Varma, A. 1986a. Optimal catalyst activity profiles in pellets. 4. Analytical evaluation of the isothermal fixed-bed reactor, *Ind. Eng. Chem. Fundam.* 25: 307–313.

Morbidelli, M., Servida, A., Carra, S., and Varma, A. 1986b. Optimal catalyst activity profiles in pellets. 5. Optimization of the isothermal fixed-bed reactor, *Ind. Eng. Chem. Fundam.* 25: 313–321.

Morbidelli, M., Brunovska, A., Wu, H., and Varma, A. 1991. Authors' reply to comments by S. Pavlou et al., *Chem. Eng. Sci.* 46: 3328–3329.

Moss, R. L., Pope, D., Davis, B. J., and Edwards, D. H. 1979. The structure and activity of supported metal catalysts. VIII. Chemisorption and benzene hydrogenation on palladium/silica catalysts, *J. Catal.* 58: 206–219.

Moyers, C. G., and Baldwin, G. W. 1997. Psychrometry, evaporative cooling and solids drying. In *Perry's Chemical Engineers' Handbook*, 7th ed., eds. Perry, R. H., Green, D. W., and Maloney, J. O., Chapter 12. New York: McGraw-Hill.

Mulcahy, F. M., Houalla, M., and Hercules, D. M. 1987. The effect of the isoelectric point on the adsorption of molybdates on fluoride modified aluminas, *J. Catal.* 106: 210–215.

Mulcahy, F. M., Fay, M. J., Proctor, A., Houalla, M., and Hercules, D. M. 1990. The adsorption of metal oxyanions on alumina, *J. Catal.* 124: 231–240.

Naoki, B., Katsuyuki, O., and Shigeki, S. 1996. Numerical approach for improving the conversion characteristics of exhaust catalysts under warming-up condition, SAE Paper 962076.

Narsimhan, G. 1976. Catalyst dilution as a means to establish an optimum temperature profile, *Ind. Eng. Chem. Proc. Des. Dev.* 15: 302–307.

Neimark, A. V., Fenelonov, V. B., and Heifets, L. I. 1976. Analysis of the drying stage in the technology of supported catalysts, *React. Kinet. Catal. Lett.* 5: 67–72.

Neimark, A. V., Kheifez, L. I., and Fenelonov, V. B. 1981. Theory of preparation of supported catalysts. *Ind. Eng. Chem. Prod. Res. Dev.* 20: 439–450.

Nicolescu, I. V., Parvulescu, V., Parvulescu, V., and Angelescu, E. 1990. Preparation of the metallic supported catalysts from precursors with different immobilization capacities on γ-Al_2O_3 surface, *Rev. Roum. Chim.* 35: 145–159.

Nyström, M. 1978. Effectiveness factors for non-uniform catalytic activity of a spherical pellet, *Chem. Eng. Sci.* 33: 379–382.

Ochoa, O., Galiasso, R., and Andreu, P. 1979. Study of some variables involved in the preparation of impregnated catalysts for the hydrotreating of heavy oils. In *Preparation of Catalysts II*, eds. Delmon, B., Grange, P., Jacobs, P., and Poncelet, G., pp. 493–506. Amsterdam: Elsevier.

Oh, S. H., and Cavendish, J. C. 1982. Transients of monolithic catalytic converters: Response to step changes in feedstream temperature as related to controlling automobile emissions, *Ind. Eng. Chem. Prod. Res. Dev.* 21: 29–37.

Oh, S. H., and Cavendish, J. C. 1983. Design aspects of poison-resistant automobile monolithic catalysts, *Ind. Eng. Chem. Prod. Res. Dev.* 22: 509–518.

Ollis, D. F. 1972. Diffusion infuences in denaturable insolubilized enzyme catalysts, *Biotechnol. and Bioeng.* 24: 871–884.

Olsbye, U., Wendelbo, R., and Akporiaye, D. 1997. Study of Pt/alumina catalysts preparation, *Appl. Catal. A* 152: 127–141.

Otsuka, M., Fujiyama, T., Tsukamoto, Y., Tsuiki, H., and Ueno, A. 1987. Control of the impregnation profile of Co in an Al_2O_3 sphere, *Bull. Chem. Soc. Japan* 60: 2881–2885.

Ott, R. J., and Baiker, A. 1983. Impregnation of γ-alumina with copper chloride. Equilibrium behaviour, impregnation profiles and immobilization kinetics. In *Preparation of Catalysts III*, eds. Poncelet, G., Grange P., and Jacobs, P. A., pp. 685–696. Amsterdam: Elsevier.

Papa, J., and Shah, Y. T. 1992. An approximate solution for effectiveness factor for a square wave shell catalyst, *Chem. Eng. Commun.* 113: 55–61.

Papadakis, V. G., Pliangos, C. A., Yentekakis, I. V., Verykios, X. E., and Vayenas, C. G., 1996. Development of high performance, Pd-based, three way catalysts, *Catal. Today* 29: 71–75.

Papageorgiou, P. 1984. Preparation of Pt/γ-Al_2O_3 pellets with internal step-distribution of catalyst, MS Thesis, University of Notre Dame.

Papageorgiou, P., Price, D. M., Gavriilidis, A., and Varma, A. 1996. Preparation of Pt/γ-Al_2O_3 pellets with internal step-distribution of catalyst: Experiments and theory, *J. Catal.* 158: 439–451.

Parfitt, G. D. 1976. Surface chemistry of oxides, *Pure Appl. Chem.* 48: 415–418.

Park, S. H., Lee, S. B., and Ryu, D. D. Y. 1981. Design of a nonuniformly distributed biocatalyst, *Biotechnol. and Bioeng.* 23: 2591–2600.

Parks, G. A. 1965. The isoelectric points of solid oxides, solid hydroxides, and aqueous hydroxo complex systems, *Chem. Rev.* 65: 177–198.

Pavlou, S., and Vayenas, C. G. 1990a. Optimal catalyst activity distribution in pellets for selectivity maximization in triangular nonisothermal reaction systems: Application to cases of light olefin epoxidation, *J. Catal.* 122: 389–405.

Pavlou, S., and Vayenas, C. G. 1990b. Optimal catalyst activity profile in pellets with shell-progressive poisoning: The case of fast linear kinetics, *Chem. Eng. Sci.* 45: 695–703.

Pavlou, S., Vayenas, C. G., and Dassios. G. 1991. Comments on optimal catalyst activity profiles in pellets, VIII. General nonisothermal reacting systems with arbitrary kinetics, *Chem. Eng. Sci.* 46: 3327–3328.

Pereira, C. J., Kubsh, J. E., and Hegedus, L. L. 1988. Computer-aided design of catalytic monoliths for automobile emission control, *Chem. Eng. Sci.* 43: 2087–2094.

Pernicone, N., and Traina, F. 1984. Commercial catalyst preparation, *Appl. Ind. Catal.* 3: 1–24.

Pfefferle, L. D., and Pfefferle, W. C. 1987. Catalysis in combustion, *Catal. Rev. – Sci. Eng.* 29: 219–267.

Pirkle, J. C., and Wachs, I. E. 1987. Activity profiling in catalytic reactors, *Chem. Eng. Prog.*, August, 29–34.

Podolski, W. F., and Kim, Y. G. 1974. Modeling the water-gas shift reaction, *Ind. Eng. Chem. Proc. Des. Dev.* 13: 415–421.

Polomski, R. E., and Wolf, E.E. 1978. Deactivation of a composite automobile catalyst pellet II. Bimolecular Langmuir reaction, *J. Catal.* 52: 272–279.

Post, M. F. M., and Sie, S. T. 1986. Process for the preparation of hydrocarbons, *Eur. Patent Appl.* 0174696.

Pranda, P., and Brunovska, A. 1993. Optimal catalyst pellet activity distributions for deactivating systems: Experimental study, *Chem. Eng. Sci.* 48: 3423–3430.

Price, D. M., and Varma, A. 1988. Effective diffusivity measurement through an adsorbing porous solid, *AIChE Symp. Ser.* 266: 88–96

Psyllos, A., and Philippopoulos, C. 1993. Performance of a monolithic catalytic converter used in automotive emission control: The effect of longitudinal parabolic active metal distribution, *Ind. Eng. Chem. Res.* 32: 1555–1559.

Radovich, J. M. 1985. Mass transfer effects in fermentations using immobilised whole cells, *Enzyme Microb. Technol.* 7: 2–10.

Ray, W. H. 1981. *Advanced Process Control*, New York: McGraw-Hill.

Remiarova, B., Brunovska, A., and Morbidelli, M. 1993. Sulphur poisoning of nonuniformly impregnated Pt–alumina catalysts, *Chem. Eng. Sci.* 48: 1227–1236.

Reuel, R. C., and Bartholomew, C. H. 1984. Effects of support and dispersion on the CO hydrogenation activity/selectivity properties of cobalt, *J. Catal.* 85: 78–88.

Righetto, L., Azimonti, G., Missana, T., and Bidoglio, G. 1995. The triple layer model revised, *Coll. and Surf. A* 95: 141–157.

Roth, J. F., and Reichard, T. E. 1972. Determination and effect of platinum concentration profiles in supported catalysts, *J. Res. Inst. Catal.* Hokkaido Univ. 20: 85–94.

Ruckenstein, E. 1970. The effectiveness of diluted porous catalysts, *A.I.Ch.E.J.* 16: 151–153.

Ruckenstein, E., and Dadyburjor, D. B. 1983. Sintering and redispersion in supported metal catalysts, *Rev. Chem. Eng.* 1: 251–354.

Ruckenstein, E., and Karpe, P. 1989. Control of metal distribution in supported catalysts by pH, ionic strength and coimpregnation, *Langmuir* 5: 1393–1407.

Ruckenstein, E., and Malhotra, M. L. 1976. Splitting of platinum crystallites supported on thin, nonporous alumina films, *J. Catal.* 41: 303–311.

Ruthven, D. M. 1984. *Principles of Adsorption and Adsorption Processes*, New York: Wiley.

Rutkin, D. R., and Petersen, E. E. 1979. The effect on selectivity of the macroscopic distribution of the components in a dual function catalyst, *Chem. Eng. Sci.* 34: 109–116.

Sadhukhan, P., and Petersen, E. E. 1976. Oxidation of napthalene in packed-bed reactor with catalyst activity profile: A design scheme for improved reactor stability and higher product yield, *A.I.Ch.E.J.* 22: 808–810.

Santacesaria, E., Carra, S., and Adami, I. 1977a. Adsorption of hexachloroplatinic acid on γ-alumina, *Ind. Eng. Chem. Prod. Res. Dev.* 16: 41–44.

Santacesaria, E., Gelosa, D., and Carra, S. 1977b. Basic behavior of alumina in the presence of strong acids, *Ind. Eng. Chem. Prod. Res. Dev.* 16: 45–46.

Santacesaria, E., Galli, C., and Carra, S. 1977c. Kinetic aspects of the impregnation of alumina pellets with hexachloroplatinic acid, *React. Kinet. Catal. Lett.* 6: 301–306.

Santhanam, N., Conforti, T. A., Spieker, W., and Regalbuto, J. R. 1994. Nature of metal catalyst precursors adsorbed onto oxide supports, *Catal. Today* 21: 141–156.

Saracco, G., and Speccia, V. 1994. Catalytic inorganic membrane reactors: Present experience and future opportunities, *Catal. Rev. Sci. Eng.* 36: 305–384.

Sarkany, J., and Gonzalez, R. D. 1983. Effect of pretreatment on dispersion and structure of silica- and alumina-supported Pt catalysts, *Ind. Eng. Chem. Prod. Res. Dev.* 22: 548–552.

Satterfield, C. N., Pelossof, A. A., and Sherwood, T. K. 1969. Mass transfer limitations in a trickle-bed reactor, *A.I.Ch.E.J.* 15: 226–234.

Scelza, O. A., Castro, A. A., Ardiles, D. R., and Parera, J. M. 1986. Modeling of the impregnation step to prepare supported Pt/Al$_2$O$_3$ catalysts. *Ind. Eng. Chem. Fundam.* 25: 84–88.

Schbib, N. S., Garcia, M. A., Gigola, C. E., and Errazu, A. F. 1996. Kinetics of front-end acetylene hydrogenation in ethylene production, *Ind. Eng. Chem. Res.* 35: 1496–1505.

Scheuch, S., Kamphuis, A. J., McKay, I. R., and Walls, J. R. 1996. A two stage fluidized/spouted bed for the granulation of catalyst powder. In *Proceedings of the 1996 IChemE Research Event*, pp. 997–999.

Schilling, L. B. 1994. *Catalysis and Biocatalysis Technologies. Leveraging Resources and Targeting Performance*, National Institute of Standards and Technology.

Scholten, J. J. F., and Van Montfoort, A. 1962. The determination of the free-metal surface area of palladium catalysts, *J. Catal.* 1: 85–92.

Schwarz, J. A. 1992. The adsorption/impregnation of catalytic precursors on pure and composite oxides, *Catal. Today* 15: 395–405.

Schwarz, J. A., and Heise, M. S. 1990. Preparation of metal distributions within catalyst supports IV. Multicomponent effects, *J. Coll. Interf. Sci.* 135: 461–467.

Schwarz, J. A., Ugbor, C. T., and Zhang, R. 1992. The adsorption/impregnation of Pd(II) cations on alumina, silica and their composite oxides, *J. Catal.* 138: 38–54.

Schwarz, J. A., Contescu, C., and Contescu, A. 1995. Methods for preparation of catalytic materials, *Chem. Rev.* 95: 477–510.

Seyedmonir, S. R., Plischke, J. K., Vannice, M. A., and Young, H. W. 1990. Ethylene oxidation over small silver crystallites, *J. Catal.* 123: 534–549.

Shadman-Yazdi, F., and Petersen, E. E. 1972. Changing catalyst performance by varying the distribution of active catalyst within porous supports, *Chem. Eng. Sci.* 27: 227–237.

Shah, A. M., and Regalbuto, J. R. 1994. Retardation of Pt adsorption over oxide supports at pH extremes: Oxide dissolution or high ionic strength? *Langmuir* 10: 500–504.

Sheintuch, M., Lev, O., Mendelbaum, S., and David, B. 1986. Optimal feed distribution in reactions with maximal rate, *Ind. Eng. Chem. Fundam.* 25: 228–233.

Shelley, S. 1997. Destroying emissions with catalysts, *Chem. Eng.*, July, pp. 57–60.

Sherrington, D. C., and Hodge, P. 1988. *Syntheses and Separations Using Functional Polymers*, New York: Wiley.

Shyr, Y.-S., and Ernst, W. R. 1980. Preparation of nonuniformly active catalysts, *J. Catal.* 63: 425–432.

Sivaraj, Ch., Contescu, Cr., and Schwarz, J. A. 1991. Effect of calcination temperature of alumina on the adsorption/impregnation of Pd(II) compounds, *J. Catal.* 132: 422–431.

Smirniotis, P. G., and Ruckenstein, E. 1993. The activity of composite catalysts: Theory and experiments for spherical and cylindrical pellets, *Chem. Eng. Sci.* 48: 585–593.

Smith, T. G., and Carberry, J. J. 1975. On the use of partially impregnated catalysts for yield enhancement in non-isothermal non-adiabatic fixed bed reactors, *Can. J. Chem. Eng.* 53: 347–349.

Soled, S. L., Baumgartner, J. E., Reyes, S. C., and Iglesia, E. 1995. Synthesis of eggshell cobalt catalysts by molten salt impregnation techniques. In *Preparation of Catalysts VI*, eds. Poncelet, G., Martens, J., Delmon, B., Jacobs, P. A., and Grange, P., pp. 989–997. Amsterdam: Elsevier.

Souza, G. L. M., Santos, A. C. B., Lovate, D. A., and Faro, A. C. 1989. Characterization of Ni-Mo/Al$_2$O$_3$ catalysts prepared by simultaneous impregnation of the active components, *Catal. Today* 5: 451–461.

Spanos, N., Vordonis, L., Kordulis, C., and Lycourghiotis, A. 1990a. Molybdenum–oxo species deposited on alumina by adsorption. I. Mechanism of the adsorption, *J. Catal.* 124: 301–314.

Spanos, N., Vordonis, L., Kordulis, C., Koutsoukos, P. G., and Lycourghiotis, A. 1990b. Molybdenum–oxo species deposited on alumina by adsorption. II. Regulation of the surface Mo^{VI} concentration by control of the protonated surface hydroxyls, *J. Catal.* 124: 315–323.

Spanos, N., Kordulis, C., and Lycourghiotis, A. 1991. Development of a methodology for investigating the adsorption of species containing catalytically active ions on the surface of industrial carriers. In *Preparation of Catalysts V*, eds. Poncelet, G., Jacobs, P. A., Grange, P., and Delmon, B., pp. 175–184. Amsterdam: Elsevier.

Spanos, N., Matralis, H. K., Kordulis, C., and Lycourghiotis, A. 1992. Molybdenum–oxo species deposited on titania by adsorption: Mechanism of the adsorption and characterization of the calcined species, *J. Catal.* 136: 432–445.

Spielbauer, D., Zeilinger, H., and Knoezinger, H. 1993. Adsorption of palladium–ammino-aquo complexes on γ-alumina and silica, *Langmuir* 9: 460–466.

Sporka, K., and Hanika, J. 1992. Development of cobalt–molybdenum hydrodesulfurization catalysts, *Collect. Czech. Chem. Commun.* 57: 2501–2508.

Srinivasan, R., Liu, H. C., and Weller, S. W. 1979. Sintering of shell molybdena–alumina catalysts, *J. Catal.* 57: 87–95.

Stenger, H. G., Meyer, E. C., Hepburn, J. S., and Lyman, C. E. 1988. Nitric oxide reduction using co-impregnated rhodium on an alumina Celcor honeycomb, *Chem. Eng. Sci.* 43: 2067–2072.

Stern, O. 1924. Zur Theorie der elektrolytischen Doppelschicht, *Z. Elektrochemie* 30: 508–516.

Stiles, A. B., and Koch, T. A. 1995. *Catalyst Manufacture*, 2nd ed., New York: Marcel Dekker.

Subramanian, S., and Schwarz, J. A. 1991. Adsorption of chloroplatinic acid and chloroiridic acid on composite oxides, *Langmuir* 7: 1436–1440.

Subramanian, S., Obrigkeit, D. D., Peters, C. R., and Chattha, M. S. 1992. Adsorption of palladium on γ-alumina, *J. Catal.* 138: 400–404.

Summers, J. C., and Ausen, S. A. 1978. Catalyst impregnation: Reactions of noble metal complexes with alumina, *J. Catal.* 52: 445–452.

Summers, J. C., and Hegedus, L. L. 1978. Effects of platinum and palladium impregnation on the performance and durability of automobile exhaust oxidizing catalysts, *J. Catal.* 51: 185–192.

Summers, J. C., and Hegedus, L. L. 1979. Modes of catalyst deactivation in stoichiometric automobile exhaust, *Ind. Eng. Chem. Prod. Res. Dev.* 18: 318–324.

Szegner, J. 1997. Effects of nonuniform catalyst distribution on inorganic membrane reactor performance: Experiments and theory, PhD Thesis, University of Notre Dame.

Szegner, J., Yeung, K. L., and Varma, A. 1997. Effect of catalyst distribution in a membrane reactor: Experiments and model, *A.I.Ch.E.J.* 43: 2059–2072.

Tanabe, K. 1970. *Solid Acids and Bases*, New York: Academic Press.

Taylor, K. C. 1984. Automobile catalytic converters. In *Catalysis Science and Technology*, vol. 5, eds. Anderson, J. R., and Boudart, M., pp. 119–170. Berlin: Springer-Verlag.

Thayer, A. M. 1994. Catalyst industry stresses need for partners as key to future success, *Chem. and Eng. News*, July 11, pp. 19–20.

Tosa, T., Mori, T., Fuse, N., and Chibata, I. 1969. Studies on continuous enzyme reactions. Part V. Kinetics and industrial application of aminoacylase column for continuous optical resolution of acyl-DL-amino acids, *Agr. Biol. Chem.* 33: 1047–1052.

Tronci, S., Baratti, R., and Gavriilidis, A. 1999. Catalytic converter design for minimisation of cold-start emissions, *Chem. Eng. Commun.*, 173: 53–77.

Tsotsis, T. T., Champagnie, A. M., Minet, R. G., and Liu, P. K. T. 1993. Catalytic membrane reactors. In *Computer Aided Design of Catalysts*, eds. Becker, E. R., and Pereira, C. J., pp. 471–552. New York: Marcel Dekker.

Uemura, Y., and Hatate, Y. 1989. Effect of nickel concentration profile on selectivity of acetylene hydrogenation, *J. Chem. Eng. Japan* 22: 287–291.

Uemura, Y., Hatate, Y., and Ikari, A. 1986. Effects of post-impregnation drying conditions on physical properties and overall reaction rate of nickel/alumina catalysts, *J. Chem. Eng. Japan* 19: 560–566.

Uemura, Y., Hatate, Y., and Ikari, A. 1987. Formation of nickel concentration profile in nickel/alumina catalyst during post-impregnation drying, *J. Chem. Eng. Japan* 20: 117–123.

Underwood, R. P., and Bell, A. T. 1987. Influence of particle size on carbon monoxide hydrogenation over silica and lanthana supported rhodium, *Appl. Catal.* 34: 289–310.

Uzio, D., Giroir-Fendler, A., Lieto, J., and Dalmon, J. A. 1991. Catalytic membrane reactors: Preparation and characterization of an active Pt/alumina membrane, *Key Eng. Mater.* 61: 111–116.

Van den Berg, G. H., and Rijnten, H. Th. 1979. The impregnation and drying step in catalyst manufacturing. In *Preparation of Catalysts II*, eds. Delmon, B., Grange, P., Jacobs, P., and Poncelet, G., pp. 265–277, Elsevier, Amsterdam.

Van Erp, W. A., Nanne, J. M., and Post, M. F. M. 1986. Catalyst preparation, Eur. Patent Appl. 0178008.

Van Herwijnen, T., Van Doesburg, H., and De Jong, W. A. 1973. Kinetics of the methanation of CO and CO_2 on a nickel catalyst, *J. Catal.* 28: 391–402.

Van Tiep, L., Bureau-Tardy, M., Bugli, G., Djega-Mariadassou, G., Che, M., and Bond, G. C. 1986. The effect of reduction conditions on the chloride content of Ir/TiO_2 catalysts and their activity for benzene hydrogenation, *J. Catal.* 99: 449–460.

Varghese, P., and Wolf, E. E. 1980. Effectiveness and deactivation of a diluted catalyst pellet, *A.I.Ch.E.J.* 26: 55–60.

Vayenas, C. G., and Pavlou, S. 1987a. Optimal catalyst activity distribution and generalized effectiveness factors in pellets: Single reactions with arbitrary kinetics, *Chem. Eng. Sci.* 42: 2633–2645.

Vayenas, C. G., and Pavlou, S. 1987b. Optimal catalyst distribution for selectivity maximization in pellets: Parallel and consecutive reactions, *Chem. Eng. Sci.* 42: 1655–1666.

Vayenas, C. G., and Pavlou, S. 1988. Optimal catalyst distribution for selectivity maximization in nonisothermal pellets: The case of parallel reactions, *Chem. Eng. Sci.* 43: 2729–2740.

Vayenas, C. G., and Verykios, X. E. 1989. Optimisation of catalytic activity distributions in porous pellets. In *Handbook of Heat and Mass Transfer*, vol. 3, ed. Cheremisinoff, N. P., pp. 135–181. Houston: Gulf Publishing Company.

Vayenas, C. G., Pavlou, S., and Pappas, A. D. 1989. Optimal catalyst distribution for selectivity maximization in nonisothermal pellets: The case of consecutive reactions, *Chem. Eng. Sci.* 44: 133–145.

Vayenas, C. G., Verykios, X. E., Yentekakis, I. B., Papadakis, E. G., and Pliangos, C. A. 1994. New three-way catalysts with Pt, Rh and Pd, each supported on a separate support, Eur. Patent Appl. 0665047A1.

Vazquez, P. G., Caceres, C. V., Blanco, M. N., and Thomas, H. J. 1993. Influence of support on concentration profiles in alumina pellets impregnated with molybdenum solutions, *Int. Commun. Heat Mass Transfer* 20: 631–642.

Verykios, X. E., Kluck, R. W., and Johnson, D. L. 1983. Fixed-bed reactor simulation with nonuniformly activated catalyst pellets. In *Modeling and Simulation in Engineering*, eds. Ames, W. F., and Vichnevetsky, R., pp. 3–10. Amsterdam: North-Holland.

Vieth, W. R., Mendiratta, A. K., Mogensen, A. O., Saini, R., and Venkatasubramanian, K. 1973. Mass transfer and biochemical reaction in enzyme membrane reactor systems – I. Single enzyme reactions, *Chem. Eng. Sci.* 28: 1013–1020.

Villadsen, J. 1976. The effectiveness factor for an isothermal pellet with decreasing activity towards the pellet surface, *Chem. Eng. Sci.* 31: 1212–1213.

Villadsen, J., and Michelsen, M. L. 1978. *Solution of Differential Equation Models by Polynomial Approximation*. Englewood Cliffs, NJ: Prentice-Hall.

Vincent, R. C., and Merrill, R. P. 1974. Concentration profiles in impregnation of porous catalysts, *J. Catal.* 35: 206–217.

Voltz, S. E., Morgan, C. R., Liederman, D., and Jacob, S. M. 1973. Kinetic study of carbon monoxide and propylene oxidation on platinum catalysts, *Ind. Eng. Chem. Prod. Res. Dev.* 12: 294–301.

Vordonis, L., Koutsoukos, P. G., and Lycourghiotis, A. 1984. Regulation of the point of zero charge and surface acidity constants of γ-Al_2O_3 using sodium and fluoride ions as modifiers, *J. Chem. Soc. Chem. Commun.* 1309–1310.

Vordonis, L., Koutsoukos, P. G., and Lycourghiotis, A. 1986a. Development of carriers with controlled concentration of charged surface groups in aqueous solutions. I. Modification of γ-Al_2O_3 with various amounts of sodium ions, *J. Catal.* 98: 296–307.

Vordonis, L., Koutsoukos, P. G., and Lycourghiotis, A. 1986b. Development of carriers with controlled concentration of charged surface groups in aqueous solutions. II. Modification of γ-Al_2O_3 with various amounts of lithium and fluoride ions, *J. Catal.* 101: 186–194.

Vordonis, L., Koutsoukos, P. G., and Lycourghiotis, A. 1990. Adsorption of molybdates on doped γ-aluminas in alkaline solutions, *Coll. and Surf.* 50: 353–361.

Vordonis, L., Spanos, N., Koutsoukos, P. G., and Lycourghiotis, A. 1992. Mechanism of adsorption of Co^{2+} and Ni^{2+} ions on the pure and fluorinated γ-alumina/electrolyte solution interface, *Langmuir* 8: 1736–1743.

Wang, J. B., and Varma, A. 1978. Effectiveness factors for pellets with step-distribution of catalyst, *Chem. Eng. Sci.* 33: 1549–1552.

Wang, L., and Hall, W. K. 1980. On the genesis of molybdena–alumina catalyst, *J. Catal.* 66: 251–255.

Wang, L., and Hall, W. K. 1982. The preparation and genesis of molybdena–alumina and related catalyst systems, *J. Catal.* 77: 232–241.

Wang, T., and Schmidt, L. D. 1981. Intraparticle redispersion of Rh and Pt–Rh particles on SiO_2 and Al_2O_3 by oxidation–reduction cycling, *J. Catal.* 70: 187–197.

Wanke, S. E., and Flynn, P. C. 1975. The sintering of supported metal catalysts, *Catal. Rev. – Sci. Eng.* 12: 93–135.

Wei, J., and Becker, E. R. 1975. The optimum distribution of catalytic material on support layers in automotive catalysis, *Adv. Chem. Ser.* 143: 116–132.

Westall, J. C. 1986. Reactions at the oxide–solution interface: Chemical and electrostatic models, *ACS Symp. Ser.* 323: 54–78.

Westall, J., and Hohl, H. 1980. A comparison of electrostatic models for the oxide/solution interface, *Adv. Coll. Interf. Sci.* 12: 265–294.

White, C. W., III, and Dadyburjor, D. B. 1989. Effect of nonuniform deactivation on the optimum design of composite cracking catalysts, *Chem. Eng. Commun.* 86: 113–124.

Widdecke, H., Klein, J., and Haupt, U. 1986. The influence of matrix structure on the activity and selectivity of polymer supported catalysts, *Macromol. Chem., Macromol. Symp.* 4: 145–155.

Wilson, G. R., and Hall, W. K. 1970. Studies of the hydrogen held by solids. XVIII. Hydrogen and oxygen chemisorption on alumina- and zeolite-supported platinum, *J. Catal.* 17: 190–206.

Wolf, E. E. 1977. Activity and lifetime of a composite automobile catalyst pellet, *J. Catal.* 47: 85–91.

Wu, C. H., and Hammerle, R. H. 1983. Development of low cost, thermally stable, monolithic three-way catalyst system, *Ind. Eng. Chem. Prod. Res. Dev.* 22: 559–565.

Wu, H. 1994. Application of optimal catalyst activity distribution theory, *A.I.Ch.E.J.* 40: 2060–2064.

Wu, H. 1995. Optimal feed distribution in isothermal fixed-bed reactors for parallel reactions, *Chem. Eng. Sci.* 50: 441–451.

Wu, H., Yuan, Q., and Zhu, B. 1988. An experimental investigation of optimal active catalyst distribution in nonisothermal pellets, *Ind. Eng. Chem. Res.* 27: 1169–1174.

Wu, H., Brunovska, A., Morbidelli, M., and Varma, A. 1990a. Optimal catalyst activity profiles in pellets VIII. General nonisothermal reacting systems with arbitrary kinetics, *Chem. Eng. Sci.* 45: 1855–1862; 46: 3328–3329.

Wu, H., Yuan, Q., and Zhu, B. 1990b. An experimental study of optimal active catalyst distribution in pellets for maximum selectivity, *Ind. Eng. Chem. Res.* 29: 1771–1776.

Xidong, H., Yongnian, Y., and Jiayu, Z. 1988. Influence of soluble aluminium on the state and surface properties of platinum in a series of reduced platinum/alumina catalysts, *Appl. Catal.* 40: 291–313.

Xue, J., Huang, Y.-J., and Schwarz, J. A. 1988. Interaction between iridium and platinum precursors in the preparation of iridium–platinum catalysts, *Appl. Catal.* 42: 61–76.

Yates, D. E., Levine, S., and Healy, T. W. 1974. Site-binding model of the electrical double layer at the oxide/water interface, *J. Chem. Soc. Faraday Trans. I* 70: 1807–1818.

Ye, J., and Yuan, Q. 1992. Optimal catalyst activity distribution for effectiveness factor, productivity and selectivity maximization in generalized nonisothermal reacting systems with arbitrary kinetics, *Chem. Eng. Sci.* 47: 615–621.

Yeung, K. L., Aravind, R., Zawada, R. J. X., Szegner, J., Cao, G., and Varma, A. 1994. Nonuniform catalyst distribution for inorganic membrane reactors: Theoretical considerations and preparation techniques, *Chem. Eng. Sci.* 49: 4823–4838.

Yeung, K. L., Aravind, R., Szegner, J., and Varma, A. 1996. Metal composite membranes: Synthesis, characterisation and reaction studies, *Stud. Surf. Sci. Catal.*, 101: 1349–1358.

Yeung, K. L., Sebastian, J. M., and Varma, A. 1997. Mesoporous alumina membranes: Synthesis, characterization, thermal stability and nonuniform distribution of catalyst, *J. Membr. Sci.* 131: 9–28.

Yeung, K. L., Gavriilidis, A., Varma, A., and Bhasin, M. M. 1998. Effects of 1,2-dichloroethane addition on the optimal silver catalyst distribution in pellets for epoxidation of ethylene, *J. Catal.* 174: 1–12.

Yortsos, Y. C., and Tsotsis, T. T. 1981. On the relationship between the effectiveness factor for the Robin and Dirichlet problem for a catalyst with variable catalytic activity, *Chem. Eng. Sci.* 36: 1734–1736.

Yortsos, Y. C., and Tsotsis, T. T. 1982a. Asymptotic behavior of the effectiveness factor for variable activity catalysts, *Chem. Eng. Sci.* 37: 237–243.

Yortsos, Y. C., and Tsotsis, T. T. 1982b. On the relationship between the effectiveness factors for the Robin and Dirichlet problems for a catalyst with nonuniform catalytic activity. The case of generalized isothermal and nonisothermal kinetics, *Chem. Eng. Sci.* 37: 1436–1437.

Zaki, M. I., Mansour, S. A. A., Taha, F., and Mekhemer, G. A. H. 1992. Chromia on silica and alumina catalysts: Surface structural consequences of interfacial events in the impregnation course of aquated Cr(III) ions, *Langmuir* 8: 727–732.

Zaman, J., and Chakma, A. 1994. Inorganic membrane reactors, *J. Membr. Sci.*, 92: 1–28.

Zhang, R., and Schwarz, J. A. 1992. Design of inhomogeneous metal distributions within catalyst particles, *Appl. Catal. A.* 91: 57–65.

Zhukov, A. N. 1996. Dependences of the point of zero charge and the isoelectric point of amphoteric solid surface on concentration and degree of binding of the ions of the background electrolyte. The case of 1 : 1 electrolyte, *Coll. J. (Engl. Ed.)* 58: 270–272.

Author Index

Subject Index